MARINE PLANTS OF THE TEXAS COAST

**Harte Research Institute
for Gulf of Mexico Studies Series**

 Sponsored by the Harte Research Institute
for Gulf of Mexico Studies,
Texas A&M University–Corpus Christi
JOHN W. TUNNELL JR., *General Editor*

A list of titles in this series is available
at the back of the book.

MARINE PLANTS
of the TEXAS COAST

Roy L. Lehman

Texas A&M University Press   *College Station*

Copyright © 2013 by Roy L. Lehman
Manufactured in China
All rights reserved
First edition
This paper meets the requirements of ANSI/NISO Z39.48 1992.
(Permanence of Paper)
Binding materials have been chosen for durability.
∞ ♻

Library of Congress Cataloging-in-Publication Data
Lehman, Roy L., 1952–
Marine plants of the Texas coast / Roy L. Lehman — 1st ed.
p. cm. — (Harte Research Institute for Gulf of Mexico Studies series)
Includes bibliographical references and index.
ISBN 978-1-62349-016-4 (flexbound : alk. paper)
ISBN 978-1-62349-089-8 (e-book)
1. Marine plants—Texas—Gulf Coast—Identification. 2. Coastal plants—Texas—Gulf Coast—Identification. I. Title. II. Series: Harte Research Institute for Gulf of Mexico Studies series.
QK188.L435 2013
579.'17709764—dc23
2013017809

All photographs are by the author.

This book is dedicated to the many mentors and friends who inspired and encouraged my studies of the Gulf of Mexico and Caribbean plants. You know who you are!

In memory of Ruth Schott O'Brien, October 13, 1921–December 26, 2012

Contents

	Preface	ix
	Acknowledgments	xi
	Introduction	1
I.	**Common Shoreline Plants of the Texas Coast**	7
	Introduction to Shoreline Plants 9	
	List of Shoreline Plants 11	
	Key to Shoreline Plant Families 13	
	Plant Descriptions and Images 14	
II.	**Seagrasses of the Texas Coast**	57
	Introduction to Seagrasses 59	
	List of Seagrasses 60	
	Key to the Seagrasses 60	
	Plant Descriptions and Images 60	
III.	**Mangroves of the Texas Coast**	73
	Introduction to Mangroves 75	
	List of Mangroves 75	
	Key to the Mangroves 75	
	Plant Descriptions and Images 76	
IV.	**Common Seaweeds of the Texas Coast**	83
	Introduction to Seaweeds 85	
	Key to the Divisions of Seaweeds 88	
	Introduction to the Common Red Seaweeds 88	
	List of Common Red Seaweeds 91	
	Generic Key to the Common Red Seaweeds 93	
	Plant Descriptions and Images 95	
	Introduction to the Common Brown Seaweeds 138	
	List of Common Brown Seaweeds 138	

Generic Key to the Common Brown Seaweeds   139
Plant Descriptions and Images   140
Introduction to the Common Green Seaweeds   153
List of Common Green Seaweeds   154
Generic Key to the Common Green Seaweeds   155
Plant Descriptions and Images   156

Appendix: Collection and Preparation Techniques	179
Glossary	185
Bibliography	199
Index	203

Preface

The book *Marine Botany* by Clinton J. Dawes has been used as the lecture textbook in marine botany and other classes at Texas A&M University–Corpus Christi. However, Dawes's book focused on the plants of tropical Florida and did not include a comprehensive coverage of the Texas plants. So the lecture and laboratory treatment of the subject has been without an acceptable text for some time, and much of the material found in this book (primarily as PowerPoint slides) has substituted for a bound volume. Students and others expressed the need for a book that consisted of information, images, and keys for the marine plants of the Texas coast. Especially lacking was a straightforward, up-to-date coverage of the macroalgae or seaweeds. In 2010, I began to organize and compile new data, revised the nomenclature, added species newly documented in the area, acquired new images, and constructed new, user-friendly keys. Several key references (besides personal observations) provided information for this work, including *Plants of the Texas Coastal Bend* (Roy L. Lehman, Ruth O'Brien, and Tammy White, 2005); *Aquatic and Wetland Plants of the Western Gulf Coast* (Charles D. Stutzenbaker, 1999); and *Illustrated Guide to the Seaweeds and Sea Grasses in the Vicinity of Port Aransas,* a scientific treatment of the Port Aransas algae (Peter Edward, 1970). Many additional references are listed in the bibliography.

Acknowledgments

Numerous individuals, groups, and institutions supported and encouraged me to complete this work. I thank and am grateful to my many students, colleagues, and friends for their assistance and participation. To all of them, too numerous to list, I offer my grateful appreciation. I want to especially thank the late Ruth O'Brien (angiosperm plants), Ryan Fikes, Veronica Thompson, and Leah Rhyne (marine macroalgae) for their assistance in reviewing the manuscript and providing important insight into their respective group of plants. Valuable technical assistance was provided by Hudson Deyoe, Joe Kowalski (UT–Pan American), and Michael J. Wynne (University of Michigan). Numerous graduate and undergraduate students have assisted by testing the keys in the laboratory/field and provided an insight into how the mind perceives the sometimes technical aspects of scientific terminology and keying procedures. Texas A&M University–Corpus Christi granted a sabbatical from teaching during the spring of 2011, which provided time to complete additional research and organization of the book.

Professionals, specialists, or phycologists may find some information with which they do not agree. I ask only that they remember that the guide was not necessarily written for them but for students and amateur naturalists.

Many thanks are due to the staff, especially Shannon Davies, at Texas A&M University Press, for putting this guide together. I also wish to thank the reviewers who provided thoughtful insight into the manuscript. In addition, to general editor of the Gulf Coast Studies, John (Wes) Tunnell Jr., who first introduced me to the coral reefs and their algae, a personal thank you.

I also wish to thank my wife, Robbie, who has spent the last 40 years keeping me company as I collected algae and other plants in the United States and Mexico. In addition to taking photographs, pressing plants, and proofreading the manuscript, she has provided friendship, enthusiasm, and joyful encouragement throughout this endeavor.

MARINE PLANTS OF THE TEXAS COAST

Introduction

It has been my intention to provide students, nature lovers, amateur naturalists, conservation workers, parks and wildlife personnel, teachers, and others with an up-to-date, easy-to-use account of the marine plants known to grow along the Texas coast and, at the same time, provide logical keys for identifying the plants. The plant groups include the common flowering plants that live along the marine shoreline (halophytes = salt-tolerant or salt-loving plants), seagrasses that live in the shallow marine waters, mangroves of the Gulf of Mexico, and the three major groups of seaweeds (marine macroalgae). A total of 52 families, 81 genera, and 140 species are included.

The book is divided into four major sections, each reflecting a different plant type as just described. The group of marine macroalgae is further divided into the three primary groups of seaweeds loosely based upon the color of the plant's thallus (body): red, brown, and green. They require a different treatment and arrangement because they are not vascular plants (angiosperms). Formal taxonomic systems place some algae in the kingdom Protista, not Plantae; however, such arrangements are frequently undergoing major rethinking as new data become available. As the marine algae (seaweeds) have photosynthetic pigments that transform sunlight into usable food energy (photosynthesis) and perform the ecological role usually credited to plants, I have chosen to use the common terms "marine plant," "seaweed," and "macroalga" interchangeably. Information on the description of the systematic arrangement is located in the introduction to the seaweeds section. The first three major sections are arranged according to the following scheme.

The halophytes, seagrasses, and mangroves species are arranged alphabetically under the names of the genera to which they belong. The genera are also arranged in this manner within their respective families; and the families are likewise under the names of the classes to which they belong. The class arrangement follows Cronquist's System of Classification.

Scientific nomenclature follows the *Vascular Plants of Texas: A Comprehensive Checklist including Synonymy, Bibliography, and Index* (1997) by Jones, Wipff, and Montgomery. A second scientific name will sometimes follow in brackets. This name was used in early floras and is included simply for the benefit of those who may already know the plant by its former name. If a common name is available for the plant, it is also listed. A guide to the pronunciation of the scientific name is located under the name. A short characterization of the plant in terms of its appearance, size, behavior, and habit follows. A species may thus be characterized as a "prostrate annual with stems to 1 foot (30 cm) or longer" or as a "submersed marine plant of shallow waters." A subsequent brief description of the thallus (seaweeds), leaves, and flowers (angiosperms), with measurements in both English and metric units is provided. Following the description is a statement on frequency of occurrence (common, occasional, etc.) and usually a note on the preferential soil type of the plant and/or nature of the site on which it is most likely to be found (zone, etc.). The flowering period is given, based largely on personal observations over a period of many years. A flowering period of March–June indicates that the plant generally, or at least sometimes, flowers as early as March and as late as June. When there is much uncertainty as to the flowering period, as in the case of rare or little-observed species, an approximation based on available data is given.

How to Use the Keys

The keys were developed specifically for identifying common plants from the Texas coast and may not be completely reliable if used in other areas. However, most genera and many species are identical throughout the Gulf of Mexico. A specimen to be identified should have not only well-developed leaves but also open flowers, as floral characters are used extensively in the keys. In some plants, fruit structure or type is necessary to determine the species. A flowering stem or branch will usually be adequate; however, if a plant is relatively small, perhaps not more than 20 inches (50 cm) high, it is best to preserve the entire plant. Label the plant with collecting data, including information on where the plant was found and the type of environment in which it was growing.

For those without experience in the use of keys, the following example of detailed directions for keying out a specimen whose identity is already known may be helpful.

- Collect a short blooming branch of Carolina wolfberry (*Lycium carolinianum*).
- Turn to the first page of Keys to the Shoreline Plant Families. Each key consists of pairs of statements, known as couplets, that are actually pairs of choices.
- In this example, the first couplet is numbered 1. The information on the right indicates where to go next; in this case, go to the second couplet because Carolina wolfberry is not a vine.

1. Plants with climbing or sometimes trailing stems: vines. . . . .
. . . . **Convolvulaceae**
1. Plants not vines. 2
2. Plants with woody stems: trees (1). 3
2. Plants scarcely woody unless near base: herbs and subshrubs 4
3. Upland trees with scalelike leaves (2) . . . . **Tamaricaceae**
3. Intertidal and subtidal trees, marine. . . . . see **Mangrove Key**
4. Plants with spines or thorns (2) 5
4. Plants of water or land habitats that tend to stay moist: aquatic or marsh plants 6
5. Plants with leaves reduced to spines (4). . . . . **Cactaceae**
5. Plants not succulent, with stiff, woody thorns. . . . . **Solanaceae**

- In the second couplet, choose couplet 4 because Carolina wolfberry is a plant with woody stems near the base and is a subshrub. The parenthetical number (1) tells you that you used the number 1 couplet to arrive at that choice.
- Turn to couplet 4 and review the choices. Always read both choices of a couplet before making a decision. If the information just does not fit, perhaps a mistake has been made. Be patient! Sometimes you may need to start over.
- Carolina wolfberry has stiff, woody thorns and belongs to the family Solanaceae, so go to the listing for that family for further direction. If two or more genera are included under the family, then a Key to the Genera will be found at the beginning of the family section (i.e., Solanaceae).

Keying is not difficult once one becomes familiar with the mechanics. To the degree possible, characters have been chosen that are easy to distinguish with the unaided eye. In many cases, however, magnification of the flowers, fruits, seeds, and so on will be required. For this purpose a binocular microscope is ideal, but usually a 10× hand lens will suffice. Unfamiliar botanical terms are used in the keys, as well as familiar words that have a special botanical meaning. The meaning of ordinary words such as *alternate, opposite, toothed, entire, simple,* or *compound* may not be obvious or clear. For the special botanical meaning of these and other terms, a glossary is provided at the back of the book.

Limits of the Region
From the Sabine River, at the border with Louisiana, the Texas coast begins a gradual southern curve that continues to the south of Baffin Bay before taking a more southeasterly line toward the border with Mexico at the mouth of the Rio Grande. Along this shoreline, the Texas coast can be divided into three ecological zones/regions based upon the amount of precipitation (rain), fresh-

water runoff (including river inflow), Gulf of Mexico marine water inflow, and the rate of evaporation (including annual air/water temperature).

From the Sabine River and nearby Galveston Bay to the Colorado River and Matagorda Bay, the estuaries are considered to be "positive," as they have a greater amount of precipitation and runoff than evaporation. In addition, the climate is more temperate. From here, this boundary arcs southwest to the southern edge of Corpus Christi Bay. These estuaries are considered to be "neutral," with precipitation and runoff typically equal to evaporation. The climate and temperature are temperate to subtropical. Along Padre Island from the mouth of the Upper Laguna Madre to the south shore of Baffin Bay, through the Land Cut, and southward to the Lower Laguna Madre and the Rio Grande, the estuary is considered "negative," as the amount of evaporation is greater than both precipitation and runoff. The Upper Laguna Madre, which is practically landlocked, has a higher salinity than the Lower Laguna Madre,

Map of the Texas coast.

which has inflow from channels from the Gulf of Mexico at both the south (Brazos Santiago Pass) and north (Mansfield Pass) ends, resulting in an exchange of water. The climate is subtropical to tropical.

Climate

The climate of the Texas coastal region is temperate to subtropical in the sense that seasons are rather poorly differentiated and the winters are short and mild. The water of the Gulf of Mexico helps maintain a consistent annual temperature and provides a source for coastal rainfall.

The long, hot summer begins in May and extends through September. Average maximum temperatures in July vary from 91°F along the coast to greater than 98°F inland. Average July minimums range from 76°F at Corpus Christi to a degree or two higher inland. Absolute summer highs range from 100°F to 105°F along the coast to 111°F inland.

Winter lasts from December through February; the temperature drops to freezing or below about four times each year (at Corpus Christi) and somewhat more frequently inland. Average January highs vary from 65°F in the upper part of the coast to 70°F along the southern coast. Quite low temperatures occur at intervals of usually several years. A low of 10°F has been recorded at Rockport and 11°F at Corpus Christi. The average date of the first killing frost varies from December 20 along the coast to November 30 inland and of the last killing frost from January 30 along the coast to February 18 inland. Some years are without killing frost along the coast.

Spring is a short season of two months beginning in March and extending through April; fall begins in October and lasts through November. Both seasons are transitional with occasional spells of both hot and winterlike weather.

Relative humidity is rather high due to the proximity of the Gulf. It is somewhat higher in the immediate vicinity of the coast than inland, varying from 80% to 90% at 8:00 a.m. to between 50% and 60% or less in the afternoons.

TABLE 1 SUMMARY OF THE FLORA

Division	Families	Genera	Species
Rhodophyta (red algae)	14	26	52
Ochrophyta (brown algae)	6	9	12
Chlorophyta (green algae)	8	10	22
Spermatophyta			
Class Magnoliopsida (dicots)	21	27	42
Class Liliopsida (monocots)	3	9	12
Total	52	81	140

I | Common Shoreline Plants of the Texas Coast

Shoreline plants.

Introduction to Shoreline Plants

Along the shoreline fringe of Texas coastal marshes is a group of plants often overlooked or thought to be of little value. In reality, they are unique vascular plants that have had to adapt to a very hostile environment, most notably, one of higher salinity. Biologists regard these plants, which are located along estuarine creeks and intertidal shorelines, as some of the earth's most important, forming the foundation for the development and maintenance of highly productive natural marsh areas. Environmental conditions necessary for the formation of a marsh include protection from wave action, freshwater, sediment deposition, and shallow shorelines with very little slope or grade. Seasonally, the marsh vegetation decays on the mud along the shoreline, recycling nutrients, which forms the base of an extensive food web that culminates in estuarine and fisheries communities.

Most of the plants of salt marshes are referred to as halophytes (salt-loving plants) and have xerophytic adaptations (adapted to grow in low freshwater [dry] conditions) that allow them to tolerate salt conditions. Morphological and anatomical adaptations to these conditions include shallow root formation with rhizomes, erect stems that develop from the rhizomes, modifications in the leaf structure (thin and dry or thick and succulent), reduced size, and specialized structures (i.e., hydathodes, hydrocytes, etc.) that reduce the loss of water.

The dominant angiosperm monocots along the coastal shorelines are *Distichlis spicata* (salt grass), *Monanthochloë littoralis* (salt-flat grass), and *Spartina alterniflora* (smooth cordgrass). Common dicots include *Batis maritima* (saltwort), *Borrichia frutescens* (sea ox-eye daisy), *Heliotropium curassavicum* (seaside heliotrope), *Limonium carolinianum* (sea lavender), *Lycium carolinianum* (Carolina wolfberry), *Machaeranthera phyllocephala* (camphor daisy), *Salicornia bigelovii* (annual glasswort), *S. virginica* (perennial glasswort), *Sesuvium portulacastrum* (sea purslane), and *Suaeda linearis* (sea blite).

Batis is an excellent model of a fleshy, thick-leafed (succulent) halophyte that is monotypic and has small flowers that are found in conelike spikes. The plant has spreading or creeping stems that root at the tips and form extensive colonies in intertidal saline areas.

Salicornia is an example of a succulent (fleshy-thick) halophyte that is unique because it has lost the leaf blade. Photosynthetic tissue has been replaced by the wrapping of the fleshy-leaf petiole around the stem.

Spartina alterniflora, an example of a halophytic plant with a thin, dry (nonsucculent) leaf, uses modifications in the leaf structure to reduce the loss of water. The leaf has a thin cuticle (outer waxy covering), deep leaf grooves on the upper sides, and thin-walled hydathodes located at the base of the grooves. With the loss of water, the hydathodes will collapse, resulting in the curling or rolling of the leaf, and thereby reduce the leaf area exposed to the air and consequent water loss.

The major ecological roles of these plants and the marshes they form include the production of organic matter with a release of nutrients, habitat and food for animals, filtration of coastal runoff, and erosion control through soil trapping and stabilization. Tidal flats are an extremely valuable habitat and should not be forgotten during decisions concerning conservation issues. Although the gently sloping areas may appear sterile and contain little or no erect vegetation, the surface sediment and substrate are composed of a multitude of plants and animals. Cyanobacteria and certain other bacteria are capable of fixing nitrogen and are the primary source of nutrients for plants surrounding the tidal flats. When the area is inundated during high tides, nitrogen mixes into the water column and is transported into salt marshes, seagrass beds, and other wetland areas. The shallow-water flats are important nurseries for numerous species of fish, crustaceans, and other invertebrates, as well as feeding grounds for many shorebirds.

Except for freshwater ponds (dominated by *Typha latifolia*, cattail) located along the shorelines, most of the marsh plants found are members of a brackish type of marsh, which is normally situated landward of salt marshes. These marshes are subjected to inundation. The flooding may be both freshwater from rainfall and subsequent runoff or marine tidal inflow from extremely high tides. These transitional marshes create an ecotone where plants from both the salt marsh community and the freshwater marsh community meet and integrate. The result is a blending of two or more adjacent vegetation types and a more diverse community with an increased number of highly adaptable plants that tend to colonize such transitional areas.

The following descriptions have been prepared for individuals with some basic knowledge of plants. If you have difficulty understanding some of the terminology or identifying structures, obtain a general botany text from the library to help you learn the parts of a plant, especially the structure of the flower. Some effort may be required, but the rewards attained with this new knowledge will be significant. The more botanical knowledge you possess or obtain, the easier plant identifications will become.

Many of the halophytic plants common to a Texas salt marsh are described in this book. The distribution and zonation associated with those plants are dependent on the concentration of salt in the substrate and water regimens.

List of Shoreline Plants

The plant species are arranged alphabetically under the names of the genera to which they belong. The genera are also arranged in this manner within their respective families, and the families are likewise under the names of the classes to which they belong. The class arrangement follows Cronquist's System of Classification. Scientific nomenclature follows the *Vascular Plants of Texas: A Comprehensive Checklist including Synonymy, Bibliography, and Index* (1997) by Jones, Wipff, and Montgomery.

Class Magnoliopsida

AIZOACEAE Carpetweed Family
Sesuvium maritimum (Walter) Britton, Sterns & Poggenberg SEA PURSLANE
S. portulacastrum (Linnaeus) Linnaeus SEA PURSLANE
S. trianthemoides Correll SEA PURSLANE
S. verrucosum Rafinesque-Schmaltz SEA PURSLANE

APIACEAE Carrot Family
Hydrocotyle bonariensis Commerson ex Lamarck MARSH PENNYWORT
H. umbellata Linnaeus UMBRELLA PENNYWORT, PENNYWORT

ASTERACEAE Aster or Sunflower Family
Borrichia frutescens (Linnaeus) A. P. de Candolle SEA OX-EYE DAISY
Machaeranthera phyllocephala (A. P. de Candolle) Shinners [*Rayjacksonia phyllocephala*] CAMPHOR DAISY

BATACEAE Saltwort Family
Batis maritima Linnaeus SALTWORT

BORAGINACEAE Heliotrope Family
Heliotropium angiospermum Murray HELIOTROPE
H. curassavicum Linnaeus SEASIDE HELIOTROPE
H. racemosum Rose & Standley HELIOTROPE

BRASSICACEAE Mustard Family
Cakile geniculata (Robinson) Millspaugh SEA ROCKET
C. lanceolata (von Willdenow) O. E. Schulz SEA ROCKET

CACTACEAE Cactus Family
Opuntia engelmannii Salm-Dyck TEXAS PRICKLY PEAR
O. macrorhiza PLAINS PRICKLY PEAR
O. stricta SOUTHERN SPINELESS CACTUS

CHENOPODIACEAE Pigweed Family
Salicornia bigelovii Torry ANNUAL GLASSWORT
S. virginica Linnaeus PERENNIAL GLASSWORT
Suaeda conferta (Small) I. M. Johnston SEA BLITE

S. linearis (Elliott) Moquin-Tandon. SEA BLITE
CONVOLVULACEAE Morning-Glory Family
Ipomoea imperati (Vahl) Grisebach [*I. stolonifera*]
FIDDLELEAF MORNING-GLORY
I. pes-caprae (Linnaeus) R. Brown GOATFOOT MORNING-GLORY
I. sagittata Poiret SALTMARSH MORNING-GLORY
GENTIANACEAE Gentian Family
Eustoma exaltatum (Linnaeus) Salisbury ex G. Don BLUEBELL GENTIAN
ONAGRACEAE Evening Primrose Family
Calylophus serrulatus (Nuttall) Raven [*C. australis*] EVENING PRIMROSE
Oenothera drummondii Hooker BEACH EVENING PRIMROSE
PLUMBAGINACEAE Leadwort Family
Limonium carolinianum (Walter) Britton SEA LAVENDER
PRIMULACEAE Primrose Family
Samolus ebracteatus Kunth SEABEACH PIMPERNEL
SOLANACEAE Nightshade Family
Lycium carolinianum Walter CAROLINA WOLFBERRY
TAMARICACEAE Salt Cedar Family
Tamarix aphylla (Linnaeus) Karsten SALT CEDAR
T. canariensis von Willdenow SALT CEDAR
T. chinensis de Loureiro SALT CEDAR
T. gallica C. Linnaeus SALT CEDAR
T. ramosissima von Ledebour SALT CEDAR

Class Liliopsida

CYPERACEAE Sedge Family
Bolboschoenus robustus (Pursh) Soják [*Scirpus maritimus*]
SALTMARSH BULRUSH
Fimbristylis castanea (Michx.) Vahl SALTMARSH FIMBRISTYLIS
Rhynchospora colorata (Linnaeus) Pfeiffer WHITE-TOPPED SEDGE
POACEAE Grass Family
Distichlis spicata (Linnaeus) Greene SALT GRASS
Monanthochloë littoralis Engelmann SALT-FLAT GRASS
Spartina alterniflora Loisel SMOOTH CORDGRASS
S. patens (Ait.) Muhl. SALT MEADOW CORDGRASS
S. spartinae (K. von Trinius) E. Merrill ex A. Hitchcock GULF CORDGRASS
Sporobolus virginicus (Linnaeus) Kunth VIRGINIA DROPSEED
Uniola paniculata Linnaeus SEA OATS
TYPHACEAE Cattail Family
Typha domingensis Persoon. NARROW-LEAVED CATTAIL
T. latifolia Linnaeus COMMON CATTAIL

Key to Shoreline Plant Families

1. Plants with climbing or trailing stems: vines . . . . **Convolvulaceae**
1. Plants not vines 2
2. Plants with woody stems: trees (1) 3
2. Plants scarcely woody unless near base: herbs and subshrubs 4
3. Upland trees with scalelike leaves (2) . . . . **Tamaricaceae**
3. Intertidal and subtidal trees, marine . . . . see **Mangrove Key**
4. Plants with spines or thorns (2) 5
4. Plants of water or land habitats that tend to stay moist: aquatic or marsh plants 6
5. Plants succulent with leaves reduced to spines (4) . . . . **Cactaceae**
5. Plants not succulent, with stiff, woody thorns . . . . **Solanaceae**
6. Plants wholly submerged or only the flowers raised to the surface (4) . . . . see **Seagrass Key**
6. Plants partly or not at all submerged 7
7. Leaves apparently lacking or reduced to scales or sheaths (6) 8
7. Leaves well developed 11
8. Stems jointed (7) 9
8. Stems not jointed 10
9. Stems not fleshy (8) . . . . **Poaceae**
9. Stems fleshy . . . . **Chenopodiaceae**
10. Sheath usually open on only one side (8) . . . . **Typhaceae**
10. Closed basal sheath; stem triangle in cross section . . . . **Cyperaceae**
11. Leaves partly or all compound or dissected (7) 12
11. Leaves simple, entire to toothed or whorled, not dissected 13
12. Flowers umbellate (11) . . . . **Apiaceae**
12. Flowers not umbellate . . . . **Brassicaceae**
13. Leaves thick-fleshy; plants of brackish or salty soils (11) 14
13. Leaves not thick-fleshy 15
14. Flowers solitary (13) . . . . **Aizoaceae**
14. Flowers unisexual and in spikes . . . . **Bataceae**
15. Ovary superior (13) 16
15. Ovary inferior 18
16. Leaves all basal (15) . . . . **Plumbaginaceae**
16. Leaves not all basal 17
17. Flowers in unrolling spikes (16) . . . . **Boraginaceae**
17. Flowers not in spikes; corolla violet . . . . **Gentianaceae**
18. Flowers in involucrate heads (15) . . . . **Asteraceae**
18. Flowers not in heads 19
19. Corolla ⅛–⅜ inch (2–7 mm) long (18) . . . . **Primulaceae**
19. Corolla much longer; petals four; stamens eight . . . . **Onagraceae**

Plant Descriptions and Images

Class Magnoliopsida

AIZOACEAE Carpetweed Family
***SESUVIUM* Sea Purslane**
⟶ sess-OO-vee-um
S. maritimum (Walter) Britton, Sterns & Poggenberg
⟶ mer-ih-TEE-mum
S. portulacastrum (Linnaeus) Linnaeus
⟶ port-yoo-luh-KAS-trum
S. trianthemoides Correll
⟶ try-an-them-OH-id-eez
S. verrucosum Rafinesque-Schmaltz
⟶ ver-oo-KO-sum

Sesuvium is a thick and fleshy perennial with branched stems that grow flat and level with the ground, often forming extensive mats more than 2 feet (60 cm) in diameter. The leaves are smooth (glabrous), waxy, and up to 1½ inches (5 cm) long. Flowers are solitary, pink, about ½ inch (1 cm) long. Sepals have a hornlike appendage. The cone-shaped fruit is about ¼ inch (7 mm) long with black seeds that are either smooth or wrinkled and lustrous (1–1.5 mm long). The plants are commonly found in saline sand or clay along the upper edge of salt marshes that border wind-tidal flats. The seeds are eaten by waterfowl.

Key to the Species

1. Stamens five...2
1. Stamens more than five...3
2. Seeds smooth (1)......................................*S. maritimum*
2. Seeds wrinkled (rugose)..........................*S. trianthemoides*
3. Stems rooting at nodes; stems mostly trailing (1).......*S. portulacastrum*
3. Stems not rooting; branching stems mostly erect..........*S. verrucosum*

S. maritimum. An annual plant that is distinguished from *S. trianthemoides* by its smooth seeds. Occasionally found on salt flats along coastal rivers. Blooms all year.

S. portulacastrum CENICILLA. A perennial with creeping stems to 6 feet (2 m) long, often forming mats. Leaves are 1 inch (3–5 cm) long, oblanceolate, and fleshy. Sepals are ¼ inch (5–7 mm) long and lavender-purple on the inner side. A common plant found along bay and island beaches. Blooms April–December.

S. trianthemoides. Annual with prostrate or reclining stems to 16 inches (40 cm) long. Leaves are fleshy, ½–1¼ inches (1.5–3 cm) long. Tepals ¼ inch (2–2.3 mm) long, pale lavender on the inner side. Plants are fairly frequent in brackish swales, marshes, and depressions along the coast. Blooms April–December.

S. verrucosum. Perennial with ascending or sprawling stems to 18 inches (45 cm) long, commonly forming colonies anchored by rhizomes. Leaves ½–1½ inches (1.5–4 cm) long, oblanceolate, and fleshy. Tepals are ½ inch (4–9 mm) long and purple to rose-lavender on the inner side. Frequent on salty or brackish soils along the coast. Blooms March–December.

Sesuvium portulacastrum (sea purslane) habit showing thick, succulent character.

Sesuvium portulacastrum (sea purslane) pink to lavender flower with five petals.

APIACEAE Carrot Family
***HYDROCOTYLE* Pennywort**
⟶ hi-droh-KOT-ih-lee
H. bonariensis Commerson ex Lamarck
⟶ bo-nar-ee-EN-sis
H. umbellata Linnaeus
⟶ um-bell-AY-tuh

Hydrocotyle has round (orbicular or peltate) leaves that are shallowly lobed. The leaves arise vertically from a lateral, creeping stem that is up to 20 inches (50 cm) long. They are attached to the center of the blade by the petiole. The flowers are light yellow to white and are found in umbels that are either simple or compound, which is the character that separates the two species. The fruit is flattened and elliptical in cross section. The plants thrive on wet, sandy soils that may be fresh (*H. umbellata*) to salty (*H. bonariensis*).

Key to the Species
1. Flowers (inflorescence) in a compound umbel . . . . ***H. bonariensis***
1. Flowers in a simple umbel. . . . . ***H. umbellata***

H. bonariensis MARSH PENNYWORT. The plant is a perennial to 14 inches (35 cm) high with creeping stems, which root at the nodes and form extensive mats. The leaves are peltate or roundish, usually 1¼–3 inches (3–8 cm) wide. Flowers are in compound umbels. The minute petals (1–1.5 mm long) are whitish to pale yellow. The plants are common on coastal sands in swales, depressions, marshes, and other moist grounds. Blooms March–November.

H. umbellata UMBRELLA PENNYWORT. A perennial much as described above but with leaves mostly smaller, ¼–1¼ inches (1–3 cm) wide. The flowers are in simple umbels, unlike those of *H. bonariensis*. The plant is locally abundant on damp sands in swales, in depressions, and around ponds. The leaves are eaten mixed in salads. Blooms March–November.

APIACEAE

▲ *Hydrocotyle umbellata* (umbrella pennywort) habit.

▲ *Hydrocotyle umbellata* (umbrella pennywort) white to yellowish flowers (close-up) in a simple umbel.

◄ *Hydrocotyle bonariensis* (marsh pennywort) orbicular leaf.

ASTERACEAE Aster or Sunflower Family

Key to the Genera
1. Leaves have a smooth (entire) margin . . . . *Borrichia*
1. Leaves have deeply toothed margins (edges) . . . . *Machaeranthera*

BORRICHIA Sea Ox-eye Daisy
Borrichia frutescens (Linnaeus) A. P. de Candolle
⟹ bore-RICK-ee-uh fru-TESS-kenz

Sea ox-eye daisy is a semitropical plant found along the upper edge of a marsh where the salinity is usually highest. It is a perennial shrub or subshrub up to 2½ feet (75 cm) tall, which often forms colonies from rhizomes. The oppositely arranged leaves are variable in size ¾–1½ inches (2–6 cm) and shape (obovate to oblanceolate to spatulate) but are characteristically thick and succulent and have a smooth margin (edge). The flower heads are perfect and mostly solitary, 1–1½ inches (2.5–3.5 cm) wide. The outside ray flowers (15–30) are bright yellow or tinged with orange. The inside disk flowers are yellow to a brownish-yellow. The mature flower heads feel sharp to the touch. The plant may bloom year-round but normally blooms April–December.

▲ *Borrichia frutescens* (sea ox-eye daisy) habit along saline shoreline. Mature flowers are hard and stickery.

▶ *Borrichia frutescens* (sea ox-eye daisy) flower with yellow ray flower and brownish-yellow disk flower; leaf opposite with smooth margins (edge).

MACHAERANTHERA Camphor Daisy
Machaeranthera phyllocephala (A. P. de Candolle) Shinners [*Rayjacksonia phyllocephala*]
🡆 mak-ee-RANTH-er-uh fill-oh-CEF-uh-lah

Camphor daisy is an annual to 2 feet (60 cm) or taller with glandular-pubescent leaves ¾–2 inches (2–5 cm) long. They have a sticky residue and, when crushed, have the odor of camphor. The leaves are alternately arranged and have deeply toothed margins (edges). Flower heads are solitary or a few together, 1–1½ inches (2.5–4 cm) wide, usually closely surrounded (subtended) by leaves. Both the disks and ray flowers are yellow. The plant is common on brackish grounds along the coast, mostly on beaches and salty, wind-tidal flats. Blooms June–September, sometimes earlier.

▲ *Machaeranthera phyllocephala* (camphor daisy) habit.

▲ *Machaeranthera phyllocephala* (camphor daisy) alternate leaf with sticky residue (camphor odor) from glands; margins deeply toothed.

◀ *Machaeranthera phyllocephala* (camphor daisy) yellow ray and disk flowers.

BATACEAE Saltwort Family
BATIS Saltwort
Batis maritima Linnaeus
> BAY-tiss mer-ih-TEE-muh

Saltwort may be found along the muddy flats of the shoreline in areas of higher salinity. It is a yellow to pale green, sometimes woody, shrublike plant 1–2 feet (30–60 cm) tall with a strong scent and spreads on creeping stems 60 inches (150 cm) long. The stem will commonly develop roots at the nodes and tip, forming large colonies. The fleshy leaves are opposite, sessile (no petiole), recurved (linear-oblanceolate), and may be up to 1 inch (25 mm) in length. Flowers, which develop June–August, are found in short conelike spikes about ¾ inch (1 cm) long. The sepals fuse to form multiple fruits (2–8), which are attached at the base of the leaves. When *Batis* is found in areas of low elevation, near or adjacent to the water, it will grow as an upright form; when it is found in slightly higher elevations, growth takes on a creeping or running form. The plant has reportedly been used for medicinal purposes to treat ulcers and as a garnish in salads because of its salty flavor.

Batis maritima (saltwort) habit.

BATACEAE

Batis maritima (saltwort) opposite, thick and succulent, sessile (lacks petiole or leaf stalk) leaf.

Batis maritima (saltwort) female flower.

Batis maritima (saltwort) male flower.

Batis maritima (saltwort) young fruits.

Batis maritima (saltwort) mature conelike fruit.

BORAGINACEAE Heliotrope Family
HELIOTROPIUM Seaside Heliotrope
⏵ hee-lee-oh-TROH-pee-um

H. angiospermum Murray
⏵ an-jee-oh-SPER-mum

H. curassavicum Linnaeus
⏵ ku-ra-SAV-ik-um

H. racemosum Rose & Standley
⏵ ray-see-MO-sum

Heliotropium is a perennial herb or small shrub found growing from a deep rhizome. Two species grow along the Texas coast. The stems are usually creeping and may reach lengths up to 16 inches (40 cm). The leaves are oblanceolate with entire (smooth) margins, about 1½ inches (4 cm) long and ½ inch (8 mm) wide, and usually succulent or succulent-like with absent petioles (sessile). Flower heads are found terminally on the branches as coiled cymes and may be 4 inches (10 cm) long. The corolla is funnel-shaped with white, minute (1.2–3.5 mm long) petals. The habitat is sandy soils near ponds, mud- or wind-tidal flats, and similar areas throughout the coast. The plant will flower throughout most of the year.

Key to the Species
1. Flowers mostly subtended by bracts . . . . *H. racemosum*
1. Flowers not subtended by bracts 2
2. Plant succulent, usually glaucous (1). . . . . *H. curassavicum*
2. Plant not succulent; corolla lobes broad and rounded. . . *H. angiospermum*

H. angiospermum HELIOTROPE. An annual or weak perennial with either erect or reclining stems. It may reach a height of 1–2 feet (30–70 cm). The leaves are ovate or broadly lanceolate. Flowers are in unrolling racemes and have white corollas with a yellow center. The plant prefers shady loamy soils and is often found around bay shorelines. Blooms April–December.

H. curassavicum SEASIDE HELIOTROPE, COLA DE MICO. A low perennial with mostly reclining or prostrate stems to 16 inches (40 cm) or longer. Leaves are ½–2 inches (1–5 cm) long, oblanceolate, blue-green, and glaucous. Flowers are in single or paired unrolling spikes; each corolla is minute (2–3 cm), white or bluish with a yellow throat. The plants are common on bay and Gulf beaches, salt flats, and other saline places. Blooms March–November.

H. racemosum HELIOTROPE. Grayish annual to 16 inches (40 cm) high with upright or reclining stems and lanceolate leaves ¾–2 inches (1.5–4 cm) long. Flowers are found in leafy unrolling racemes. The corolla is ¼ inch (8–12 mm) wide, white with yellow throat. Found frequently on deep sands. Blooms April–December.

Heliotropium curassavicum (seaside heliotrope) habit.

Heliotropium curassavicum (seaside heliotrope) white petals of minute flowers.

Heliotropium curassavicum (seaside heliotrope) coiled paired spikes of flowers.

BRASSICACEAE Mustard Family
CAKILE Sea Rocket
    ➠ kah-KIL-ee

C. geniculata (Robinson) Millspaugh
    ➠ gen-ICK-yoo-late-uh

C. lanceolata (von Willdenow) O. E. Schulz
    ➠ lan-see-oh-LAY-tuh

Cakile is an annual herb with thick, fleshy stems and leaves. The leaves are alternately arranged and pinnatifid in shape. The cream to white to lightly lavender flower is perfect, has four petals, and is found in a raceme. The fruit is unusual with a longitudinal partition down its length.

Key to the Species
1. Petals ¼ inch (4.5–5 mm) long; fruit pod delicately four-ridged . . . . *C. geniculata*
1. Petals ¼–¾ inch (6–20 mm) long; fruit pod coarsely eight-ridged . . . . *C. lanceolata*

C. geniculata. An annual with upright or prostrate stems to 16 inches (40 cm) long. The leaves are 1–4 inches (2–10 cm) long, entire (smooth) to toothed edge or lobed and fleshy. Flowers are in racemes, the petals white or lavender. It is a common plant of Gulf of Mexico barrier island beaches. Blooms March–October.

C. lanceolata. Annual to about 20 inches (50 cm) tall with fleshy pinnatifid leaves 1–3 inches (3–7 cm) long. Flowers are racemose, fragrant, with petals of white, pink, or lavender. Texas species is ssp. ***pseudoconstricta*** Rodman. The plants are native to the Yucatán, Mexico, but have naturalized on the lower Texas coast. Blooms year-round.

Cakile geniculata (sea rocket) habit.

Cakile geniculata (sea rocket) perfect, white flower with four petals located in a raceme.

Cakile geniculata (sea rocket) succulent; alternately arranged leaves pinnatifid in shape.

CACTACEAE Cactus Family
OPUNTIA Texas Prickly Pear
Opuntia engelmannii Salm-Dyck
⏵ o-PUN-tee-a en-gel-MAN-nee-eye

Texas prickly pear stems are upright or sprawling and often form broad clumps to 6 feet (2 m) or taller. Joints (pads) are usually 6–12 inches (15–30 cm) long, mostly obovate or subcircular. Flowers are 3–4 inches (7–10 cm) long and about as wide, with the petals yellow to orange or red. The fruits are 1–2 inches (2.5–8 cm) long, obovoid or pear-shaped, red to purple, and edible. It is common on coastal dunes, flats, and shell deposits. The plant is attractive and showy when in flower. Blooms April–May, sparingly later.

NOTE: *Opuntia macrorhiza* and *O. stricta* are also found on Texas barrier islands and along saline shorelines.

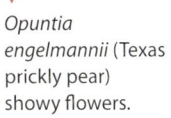
Opuntia engelmannii (Texas prickly pear) habit with red fruit.

Opuntia engelmannii (Texas prickly pear) showy flowers.

▲ *Opuntia engelmannii* (Texas prickly pear) pad with needles.

CHENOPODIACEAE Pigweed Family

Key to the Genera
1. Plant apparently leafless .... *Salicornia*
1. Plant leafy with perfect flowers .... *Suaeda*

SALICORNIA
⟹ sal-eye-KOR-nee-uh
S. bigelovii Torrey
⟹ big-eh-LOV-ee-eye
S. virginica Linnaeus
⟹ vir-JIN-ih-kuh

Key to the Species
1. Scales abruptly pointed below spikes .... *S. bigelovii*
1. Scales blunt or rounded below spikes .... *S. virginica*

The two species of *Salicornia* (commonly called glasswort) are examples of halophytic (salt-loving) plants that exhibit a thick, succulent type of leaf anatomy and are commonly found in salt marshes and flats along the edge of the water. This succulent's survival mechanism is through the accumulation of water in the plant tissues. Salinity concentrations are reduced as salts in the cells are diluted with increased amounts of water. In addition, the rapid growth of new tissues associated with this plant has a diluting effect since salts are distributed to a greater amount of tissue.

Some plant taxonomists express the view that the two species are ecomorphs and should be classified as a single species. However, the morphological characters used in identification of the two species are consistent. *Salicornia bigelovii* tolerates inundation and higher salinities (slight habitat differences). Scales of the spikes are different for each species (sharp-tipped scales below the spikes in *S. bigelovii*; blunted or rounded in *S. virginica*). *Salicornia bigelovii* is an annual whereas *S. virginica* is a perennial, and growth forms are somewhat different. *Salicornia bigelovii* is a more upright, solitary plant, whereas *S. virginica* forms extensive mats.

Both species are low-growing, erect, fleshy herbs with jointed, woody-cored primary stems. The plants have erect branches with minute flowers that are sunken in the cavities of the internodes of the spike, although the ovary still maintains a superior position. The flowers are arranged in tiny sessile three-flowered cymes sunken into the fleshy stem axis. The ovaries are superior. The flowers bloom from May through October as the proper environmental condition occurs. The oppositely arranged leaf blades are reduced to minute scales. The leaves are peculiar in that the fleshy leaf petiole (leaf base) is altered and wraps around the stem, forming a bulbous shape. Old stems lose these petioles and appear segmented and woody.

Salicornia has a narrow salinity range (3.1%) and, like most halophytes, germinates best under freshwater conditions. This most often occurs as soil salinity is reduced during the spring and winter. These halophytes can be intolerant of elevated salinity levels, especially while in the process of germination.

Geese feed on the branches of glasswort. In the fall, ducks (especially pintail) eat the stem tips when the glasswort turns a reddish color with maturity. In the past, glasswort plants were harvested and burned for potash.

S. bigelovii ANNUAL OR BIGELOW'S GLASSWORT. An annual to 2 feet (60 cm) tall with fleshy jointed stems. The spikes are fleshy, 1–5 inches (2–12 cm) long and about ¼ inch (4–7 mm) wide, with the minute flowers sunken in the spike. The plant is common in salt marshes, on tidal flats, and along

▲ *Salicornia bigelovii* (annual glasswort) stems lack leaf blades.

◀ *Salicornia bigelovii* (annual glasswort) habit.

◀◀ *Salicornia bigelovii* (annual glasswort) male flower.

◀ *Salicornia bigelovii* (annual glasswort) joint showing point below scale.

bay and island beaches. Blooms May–October, usually after a soaking rain.

S. virginica PERENNIAL OR WOODY GLASSWORT.
A perennial to 14 inches (35 cm) or taller with fleshy jointed stems, forming dense mats or clumps. The spikes are fleshy, ½–2 inches (1–5 cm) long and about ¼ inch (3–5 mm) wide, the flowers sunken in the spike. Frequent in same habitats but are located upland of *S. bigelovii*. Blooms May–September.

▲ *Salicornia bigelovii* (annual glasswort) female flower.

▶ *Salicornia virginica* (perennial glasswort) habit.

▶ *Salicornia virginica* (perennial glasswort) male and female flowers (close-up).

▶ *Salicornia virginica* (perennial glasswort) older red tips.

SUAEDA Sea Blite
⟹ soo-AY-duh
S. conferta (Small) I. M. Johnston
⟹ kon-FER-tuh
S. linearis (Elliott) Moquin-Tandon.
⟹ lin-eh-AIR-iss

Key to the Species
1. Perennial, leaves ¼–¾ inch (5–15 mm) long . . . . *S. conferta*
1. Annual, linear leaves ½–1½ inches (1–4 cm) long . . . . *S. linearis*

Suaeda linearis is a shrubby plant found along the edges of tidal flats and may form dense clumps. This annual herb has erect, profusely branched, sometimes red-tinged stems (betacyanin) with simple, entire, dark green, fleshy leaves. The leaves are usually flat on one side and, in cross section, rounded on the other. The upper leaves are reduced in size and have an alternate arrangement. During the summer and fall, small green flowers are found in terminal spikes, either singly or in clusters of three. Also found within the Texas coastal marshes is the perennial shrub *S. conferta*. It is at times difficult to distinguish from *S. linearis*. Mostly, *S. conferta* is found along the upper edge of marshes and is more erect with small, very brittle, blue-gray leaves.

NOTE: Additional species of *Suaeda* grow on barrier islands. They are found less frequently and require the use of a microscope or hand lens to confirm the species.

S. conferta. Perennial with prostrate or ascending stems to 2 feet (60 cm) long. Leaves ¼–¾ inch (5–15 mm) long by about ⅛ inch (1–3 mm) wide, fleshy, usually glaucous. Flowers in axils of reduced upper leaves. Perianth is minute (2–3 mm) long, fleshy, green. Frequently found along bay beaches or on salt flats. Blooms primarily March–December.

S. linearis. Annual to 30 inches (75 cm) high, often forming dense clumps. Leaves ½–1½ inches (1–4 cm) long, fleshy, narrowly linear. Flowers are found in spicate panicles; the greenish perianth is 1–2 mm high and fleshy. A common plant found along bay beaches or on salt flats. Blooms May–December.

◀ *Suaeda linearis* (sea blite) red (betacyanin) fruit.

▲ *Suaeda linearis* (sea blite) leaves flat on one side and convex on the other (plano-convex).

▲ *Suaeda linearis* (sea blite) flower spike.

▶ *Suaeda conferta* (sea blite) habit.

CONVOLVULACEAE Morning-Glory Family
IPOMOEA
⟶ eye-poh-MEE-a
I. imperati (Vahl) Grisebach [*I. stolonifera*]
⟶ im-per-AH-tee
I. pes-caprae (Linnaeus) R. Brown
⟶ pes-KAP-ray
I. sagittata Poiret
⟶ saj-ih-TAY-tuh

Key to the Species
1. Leaves thick-leathery 2
1. Leaves thin, sagittately lobed . . . . *I. sagittata*
2. Corolla white with yellow throat (1) . . . . *I. imperati*
2. Corolla purple . . . . *I. pes-caprae*

I. imperati FIDDLELEAF MORNING-GLORY. Perennial with trailing stems to 18 feet (6 m) or longer, rooting at the nodes. The leaves are 1–2½ inches (2–6 cm) long, entire to variously lobed, thick-leathery, and notched apically. The corolla is 1½–2½ inches (4–6 cm) long, white with a yellow throat. This showy, attractive plant is common on barrier islands and occasional along bay beaches. Blooms April–December.

▲ *Ipomea imperati* (fiddleleaf morning-glory) thick-leathery leaf with apical notch.

◄▲ *Ipomea imperati* (fiddleleaf morning-glory) white flower with yellow throat.

◄ *Ipomea imperati* (fiddleleaf morning-glory) white flower (side view).

I. pes-caprae GOATFOOT MORNING-GLORY. Perennial vine with thick trailing or slightly twining stems to 30 feet (10 m) or longer, rooting at the nodes. Leaves 2–4 inches (6–10 cm) long, about as broad, thick and leathery. Corolla 2–2½ inches (5–6 cm) long, rose-purple. Texas species is ssp. ***brasiliensis*** (Linnaeus) van Oostsroom. This attractive, showy plant is common on barrier islands, mostly along upper beaches and lower dunes, also occasional along bay beaches. Blooms May–December.

Ipomea pes-caprae (goatfoot morning-glory) habit with rose-pink flowers.

Ipomea pes-caprae (goatfoot morning-glory) leaf shaped like goat's footprint.

I. sagittata SALTMARSH MORNING-GLORY. Perennial vine with twining stems to 5 ½ feet (1.7 m) or longer. Leaves 1–4 inches (3–10 cm) long, sagittate (arrowhead-shaped). The corolla is 2¾–3¼ inches (7–8 cm) long, rose-purple, rarely white. Occasional along bay and island beaches or more frequent in coastal marshes. Flowers are usually abundant, showy, and attractive. Blooms May–September.

Ipomea sagittata (saltmarsh morning-glory) pink flower (close-up).

GENTIANACEAE Gentian Family
EUSTOMA Bluebell Gentian
Eustoma exaltatum (Linnaeus) Salisbury ex G. Don
⇒ yoo-STOH-muh eks-all-TAY-tum

Bluebell gentian is an erect annual to 2½ feet (70 cm) tall with opposite, sessile, glaucous, and ovate-oblong leaves 1½–3 inches (4–7 cm) long. Flowers are loosely cymose or solitary, the corolla lobes ½–1 inch (1.5–3 cm) long, violet to lavender with a darker splotch (purple) at base. Frequent in damp brackish or salty soils along the coast, chiefly in flats, in depressions, along beaches, and often found on protected islands along the Texas coast (i.e., Laguna Madre). Blooms May–October.

Eustoma exaltatum (bluebell gentian) habit with showy flower.

Eustoma exaltatum (bluebell gentian) opposite and sessile leaves.

Eustoma exaltatum (bluebell gentian) lavender flower with purple splotch at base.

ONAGRACEAE Evening Primrose Family

Key to the Genera and Species
1. Stigma discoid . . . . *Calylophus serrulatus*
1. Stigma narrowly four-lobed . . . . *Oenothera drummondii*

CALYLOPHUS Evening Primrose
Calylophus serrulatus (Nuttall) Raven [*C. australis*]
   ▶ kal-ih-LOF-us ser-ra-LA-tus

Evening primrose is a perennial with upright stems to about 16 inches (40 cm) tall. Leaves are linear or oblanceolate, mostly ¾–1¼ inches (1.5–3 cm) long. Flowers are axillary with a discoid stigma and have yellow petals about ½ inch (10–12 mm) long. An attractive plant found frequently on sand or shell deposits along bay beaches and barrier islands. Blooms March–December.

Calylophus serrulatus (beach primrose) four-petal flower showing discoid-shaped stigma.

OENOTHERA Beach Evening Primrose
Oenothera drummondii Hooker
⏵ ee-no-THEE-ruh drum-AWN-dee-eye

Beach evening primrose is a perennial to 30 inches (75 cm) tall with upright to spreading stems. Leaves are ½–3 inches (1–7 cm) long, mostly oblanceolate and grayish with a soft pubescence. Flowers are found in the upper axils and have a four-lobed stigma. The yellow petals are 1–1½ inches (2.5–4 cm) long. The showy and attractive flowers are common on the dunes of barrier islands and occasionally along bay beaches. Blooms March–December.

Oenothera drummondii (beach evening primrose) yellow flower showing narrowly four-lobed stigma.

Oenothera drummondii (beach evening primrose) yellow flower showing narrowly four-lobed stigma (side view).

PLUMBAGINACEAE Leadwort Family
LIMONIUM Sea Lavender
Limonium carolinianum (Walter) Britton
⟹ Lim-OH-nee-um kair-oh-lin-ee-AN-um

Sea lavender is a small (15 inches; 38–50 cm), perennial herb with woody roots and (barely) petioled, thick leaves arranged into a rosette. The flowers are on a single, tall inflorescence arising from the basal leaves and are erect, nearly naked, and branched above the middle into panicles (spikes). The corolla has five nearly distinct violet to lavender petals, with long claws; the five stamens are attached to their bases. The flowers have fine hairs, at least at the base. The sepals are mostly membranous. Blooms May–November.

This plant has several adaptations that help it survive saline inundation, including well-developed specialized tissue (aerenchyma) in the petiole and roots, typical salt glands on the leaves and stems (salt secretor), and a specialized growth form. The growth of successive rosettes of leaves elevates the plant above the normal level of the marsh soil, thereby avoiding inundation. The leaf size is dependent upon salt concentration and is smaller with increasing salinity. It is normally found in irregularly flooded salt marshes.

Limonium carolinianum (sea lavender) habit.

PLUMBAGINACEAE

▶
Limonium carolinianum (sea lavender) plant with rosette of basal leaves and a young panicle.

▼
Limonium carolinianum (sea lavender) flower (close-up).

PRIMULACEAE Primrose Family
SAMOLUS Seabeach Pimpernel
Samolus ebracteatus Kunth
⟹ sam-OH-lus e-brak-tee-AY-tus

Seabeach pimpernel is a perennial to 20 inches (50 cm) tall with spatulate (spoon-shaped) leaves 1½–5 inches (4–12 cm) long that are basal (no petiole) and are arranged alternately. Flowers are five-lobed, bell-shaped on slender spreading stalks (racemes) with the corolla (about ¼ inch [5–7 mm]) long, white aging to pink or lavender. Fruit is a round capsule that has many seeds. The plant is frequently found on damp, brackish, or salty low grounds along the coast that are often flooded. Blooms March–November. The plant resembles *Limonium carolinianum* and is found in the same habitat.

▲ *Samolus ebracteatus* (seabeach pimpernel) flower spike (raceme).

◀ *Samolus ebracteatus* (seabeach pimpernel) habit showing elevated spatulate leaves.

SOLANACEAE Nightshade Family
LYCIUM Carolina Wolfberry
Lycium carolinianum Walter
> LY-see-um kair-oh-lin-ee-AN-um

Carolina wolfberry is a spiny semi-evergreen shrub to 3 feet (1 m) or taller with upright to spreading stems. The leaves are ½–2 inches (1–5 cm) long, linear, spatulate, and thick-succulent. Flowers are axillary with the corolla ½–1 inch (1.5–2.5 cm) wide, blue-violet, or lavender. Fruits mature to a red color. Stems are armed with spines. The plant is common along the coast, mostly along beaches, on salt flats or in marshes. Blooms April–October, sometimes later. It is beneficial as a food source for birds, especially the endangered whooping crane. The juicy berries are bitter and may be eaten raw or prepared as a sauce. The plant is also winter forage for livestock.

◄ *Lycium carolinianum* (Carolina wolfberry) axillary flowers with a lavender corolla.

◄ *Lycium carolinianum* (Carolina wolfberry) fruit.

▲ *Lycium carolinianum* (Carolina wolfberry) spines along the stem.

TAMARICACEAE Salt Cedar Family
TAMARIX Salt Cedar
⟹ TAM-uh-riks

T. aphylla (Linnaeus) Karsten
⟹ a-FIL-uh

T. canariensis von Willdenow
⟹ kuh-nair-ee-EN-sis

T. chinensis de Loureiro
⟹ chi-NEN-sis

T. gallica C. Linnaeus
⟹ GAL-ee-kuh

T. ramosissima von Ledebour
⟹ ram-oh-SIS-ee-muh

Tamarix is an introduced plant from Europe that has naturalized in the southern and western United States. It is commonly found on slightly elevated sandy soils of fresh and saline wetlands. The plants are shrubs or trees with tiny leaves composed of scales scarcely more than 1/16 inch (1 mm) long. Flowers are arranged in spikelike racemes with the stamens borne on a fleshy, lobed disk.

Key to the Species
1. Leaves sheathing the stem. . . . . *T. aphylla*
1. Leaves sessile, not sheathing. 2
2. Filaments inserted between the lobes of disk (1). 3
2. Filaments not inserted as above. 4
3. Sepals finely toothed; petals obovate (2). . . . . *T. ramosissima*
3. Sepals more or less entire; petals obovate to elliptic . . . . *T. chinensis*
4. Sepals finely toothed; rachis of raceme usually papillate (2). . . . .
. . . . *T. canariensis*
4. Sepals entire or subentire; raceme rachis smooth . . . . *T. gallica*

T. aphylla ATHEL. Evergreen tree to 18 feet (6 m) or taller with gray-green leaves. Petals about 1/8 inch (2 mm) long, mauve-pink. Frequent in cultivation, sometimes used for windbreaks and screens; not naturalized. Native of North Africa and South Asia. Blooms July–September.

T. canariensis. Deciduous shrub or small tree to 8 feet (2.5 m) or taller. Petals 1/8 inch (1.2–1.5 mm) long, pink or purple-rose, falling early. Sometimes planted for screens or soil stabilization along the coast, becoming naturalized. Native of Mediterranean region and Canary Islands. Blooms May–September.

T. chinensis. Straggling deciduous shrub or small tree to 8 feet (2.5 m) or taller. Petals ⅛ inch (1.5–2.5 mm) long, pink or purplish. Planted for sand stabilization and windbreaks along the coast; also naturalized. Native of China and Japan. Blooms April–October.

T. gallica TAMARISCO, ROMPEVIENTOS. Straggling deciduous shrub or small tree to 9 feet (3 m) or taller. Petals ⅛ inch (1.5–1.7 mm) long, pink or purplish, falling early. Frequently planted along the coast for screens, windbreaks, and soil stabilization; also widely naturalized. Native of southern Europe. Blooms April–July or later.

T. ramosissima. Much like *T. gallica*, with pink petals ⅛ inch (1–1.7 mm) long. Sometimes planted for screens and windbreaks along the coast. Native of China, Russia, and South Asia. Blooms May–September.

Tamarix gallica (salt cedar) habit.

Tamarix gallica (salt cedar) flowers.

Class Liliopsida

CYPERACEAE Sedge Family

Key to the Genera
1. Achene lenticular with a tubercle at top.................*Rhynchospora*
1. Achene without a tubercle at top, often pointed......................2
2. Achene smooth (1).......................................*Fimbristylis*
2. Achene cross-ridged................................**Bolboschoenus**

BOLBOSCHOENUS Saltmarsh Bulrush
Bolboschoenus robustus (Pursh) Soják [*Scirpus maritimus*]
⟹ bulb-oh-SKEE-nus roh-BUS-tus

Saltmarsh bulrush is a perennial plant growing from rhizomes with white, brown, or reddish culms (jointed, hollow stem) to 6 feet (2 m) tall. The three-angled stem forms an erect bract that appears as a continuation of the culm. The inflorescence is 1–6 inches (3–15 cm) long and generally has a terminal cluster of 3–4 sessile spikelets with 25–50 flowers. It has about four bristles approximately as long as the seed that are retrorsely barbed. The seeds are minute (2–3 mm long), plano-convex, smooth, and shiny dark-brown at maturity. The plant is frequently found along sandy beaches within low salinity marshes. Blooms March–December.

Bolboschoenus robustus (saltmarsh bulrush) habit.

FIMBRISTYLIS Saltmarsh Fimbristylis
Fimbristylis castanea (Michx.) Vahl
⟹ fim-bree-STY-liss kas-TAN-ee-uh

Saltmarsh fimbristylis is a perennial 2–3 feet (about 1 m) tall, forming dense clumps in brackish to saline marshes. Linear leaves are thin and only 1/16 inch (1–1.5 mm) wide. Numerous small ovoid spikelets 1/4–3/4 inch (7–20 mm) long are found in a compound umbellate cyme. Achenes are small, about 1/8 inch (1.8 mm) long. Frequently found in brackish or saline sands along the coast, mostly along the shores or flats of barrier islands. Blooms May–November.

Fimbristylis castanea (saltmarsh fimbristylis) female flowers.

Fimbristylis castanea (saltmarsh fimbristylis) male flowers.

RHYNCHOSPORA White-Topped Sedge
Rhynchospora colorata (Linnaeus) Pfeiffer
⏵ rin-koh-SPOR-uh kol-uh-RAH-tuh

White-topped sedge is a colonial perennial 6–20 inches (15–50 cm) tall with small leaves ⅛ inch (2–4 mm) wide. Spikelets are small, ¼ inch (5–7 mm) long, and are found in dense white heads. A common plant found in moist, sandy depressions and swales along the coast. Blooms April–December.

Rhynchospora colorata (white-topped sedge) habit.

POACEAE Grass Family

Key to the Genera
1. Leaf blades about ¼ inch (1 cm) long; stoloniferous, mat-forming perennial with inconspicuous unisexual spikelets . . . . **Monanthochloë**
1. Leaf blades more than ¼ inch (1 cm) long; plants not as above 2
2. Inflorescence a panicle or open raceme (1) 3
2. Inflorescence a spike; spicate raceme 4
3. Primary branches not spicate (2) . . . . **Sporobolus**
3. At least some of the primary branches rebranched . . . . **Uniola**

4. Inflorescence with two (rarely one) to several primary branches (2).........
....*Spartina*
4. Inflorescence a terminal spike or spicate raceme ....**Distichlis**

DISTICHLIS Saltgrass
Distichlis spicata (Linnaeus) Greene
⟶ DIS-tik-lis spy-KAH-tuh

Saltgrass is found within high-saline areas of salt marshes and along the edges of wind-tidal flats. It is locally abundant along the southern Texas coast, often growing in dense stands within wet depressions. The low-growing dioecious perennial has erect stems and leaves that emerge from stout, scaly rhizomes. The stems are 10 inches to 2 feet (10–60 cm) long, hollow, stiff, and round. Leaves are ¼ inch (1 cm) wide and 1–2 inches (2.5–5 cm) long, numerous, short, and overlapping. They have smooth margins and when dry, are inwardly rolled, a normal response to desiccation, as specialized leaf cells called hydathodes collapse with the loss of water. The flowers are found terminally on the plants. The male spikelets have 8–12 flowers, whereas the female is usually five-flowered. Blooms June–October. The plant is often confused with *Sporobolus virginicus* (coastal dropseed), which has upper leaves that are reduced in size. Saltgrass is a good forage plant in salty sites where most other grass plants cannot survive but is more important as a species of substrate stabilization and a habitat of nesting waterfowl.

▲ *Distichlis spicata* (saltgrass) flower.

◀ *Distichlis spicata* (saltgrass) habit.

▲ Comparison of flower of *Distichlis spicata* (top) and *Sporobolus virginicus* (bottom).

▲ *Distichlis spicata* (saltgrass) numerous, short, overlapping leaves (flat plane).

MONANTHOCHLOË Salt-flat Grass
Monanthochloë littoralis Engelmann
⏩ mo-nan-THO-klow-ee lit-tor-AY-lis

Salt-flat grass is a low-growing, creeping perennial grass in poorly drained, brackish to saline flats along estuarine shores. Often found surrounded by blue-green algal mats, *Batis maritima* (saltwort) and *Salicornia virginica* (glasswort), it may be the only grass found on salt-, mud-, or wind-tidal flats. The plants are typically located in the upper regions of these areas where it may grow into extensive colonies that emerge from the hard rhizomes and are spread by stolons. The thick mat formed from the erect, wiry stems (up to 10 inches [25 cm] tall) have short (¼–¾ inch [5–10 mm] long), stiff, grayish-green, overlapping leaves. The inconspicuous male and female flowers, which emerge in the spring, are hidden in the terminal leaf sheaths of separate plants (dioecious) and are difficult to find. The plant is of poor grazing value but is important for its capacity to aid in the stabilization of marsh areas and low (landward edge) tidal flats. Blooms during warm seasons.

Monanthochloë littoralis (salt-flat grass) habit.

▲ *Monanthochloë littoralis* (salt-flat grass) flowering.

◀ *Monanthochloë littoralis* (salt-flat grass) female flower (close-up).

◀ *Monanthochloë littoralis* (salt-flat grass) male flower (close-up).

SPARTINA Cordgrass
⟹ spar-TEE-nuh
S. alterniflora Loisel
⟹ al-tern-ih-FLOR-uh
S. patens (Ait.) Muhl.
⟹ PAT-enz
S. spartinae (K. von Trinius) E. Merrill ex A. Hitchcock
⟹ spar-TIN-ay

Key to the Genera
1. Plant tufted; blades with needlelike tips . . . . *S. spartinae*
1. Plant emerging from rhizomes; blades lack needlelike tips 2
2. Spikelets ¼ inch (10 mm); lower glumes ⅛ inch (3–5 mm); lemmas ⅛–¼ inch (3–10 mm) (2) . . . . *S. patens*
2. Spikelets ⅜ inch (10 mm); lower glumes to ⅜ inch; lemmas to >⅜ inch . . . . *S. alterniflora*

S. alterniflora SMOOTH CORDGRASS, SALTWATER CORDGRASS. A perennial with robust culms and elongating rhizomes. The height of the spreading culms may vary but is typically uniform within stands, to 6 feet (2 m) tall with leaf blades to 16 inches long and ½ inch wide (40 cm × 10 mm) that are smooth and tapered. The margin of the sheath is puberulent and has a ligule with a ring of trichomes. The inflorescence is a panicle with spikes numbering 5–30 per panicle, 2–4 inches (4–10 cm) long, ⅛ inch (3–5 mm) thick, that are tightly appressed and overlapping or diverging slightly. The spikelets are sessile, 10–50 per spike, about ¼ inch (10 mm) long, and are crowded on one side of the panicle branch. The second glume is as long as the spikelet, and the first glume is shorter. Keels of the glumes and lemma are slightly pubescent. The fruit is an olivaceous to yellow grain that is linear-ellipsoid in shape and about ¼ inch (7 mm) long. This is an abundant, important salt marsh plant. It generally occurs as extensive stands in saline areas or is intermixed with other species in brackish areas. Blooms spring–fall.

Spartina alterniflora (smooth cordgrass) habit.

S. patens SALT MEADOW CORDGRASS. A wiry perennial that spreads from elongating rhizomes of small diameter. The culms are spreading or erect, to 3–4 feet (1 m) tall. The mostly involute (rolled inward) leaves are slender to 16 inches long and ¼ inch wide (40 cm × 3 mm). The inflorescence is a panicle 3½–8 inches (9–20 cm) long with spikes 2–7 per panicle. The spikes are ½–2¾ inches (1–7 cm) long, usually diverging at angles of 10–45 degrees. The spikelets number 2–50 per spike and are ¼–½ inch (5–15 mm) long. The glumes are unequal in size. Both the glume and lemma are rough and bristly on the keels, at least on the distal aspect. The floret is perfect. The fruit is a flat to ellipsoidal, olivaceous grain. Generally, it is found abundantly in dense stands on higher elevations of saline marsh in mostly saturated saline clay, especially near the edge of upland areas. Also may be intermixed with other plants in brackish marshes. Blooms summer–fall.

S. spartinae GULF CORDGRASS. A perennial that forms dense clumps that arise from a nonrhizome root system. The leaf blades are short, stiff, and inrolled on drying and have a sharp point. The flowers are stout, 6–10 inches (15–25 cm) long with 10–75 spikes per head, each ⅜–1½ inches (10–35 mm) long. Gulf cordgrass is found on fresh to brackish coastal flats and sometimes along the upland edges of salt marshes. It tolerates short periods of submergence but prefers the transitional zone between wetland and upland areas. The plant grows in dense clumps and provides habitat for many small animals. Blooms spring–summer.

Spartina spartinae (gulf cordgrass) habit.

SPOROBOLUS Virginia Dropseed
Sporobolus virginicus (Linnaeus) Kunth
→ spoh-ROB-oh-lus vir-JIN-ih-kus

Virginia dropseed is a low perennial to 15 inches (38 cm) tall with creeping rhizomes. The firm, closely crowded leaves (two-ranked) have conspicuously overlapping sheaths and short blades. The panicles are short, ¾–4 inches (2–10 cm) long, and contracted. The fruits are free from the lemma and palea and will fall from the mature spikelet, resulting in its common name, dropseed. Locally frequent in sandy soils along saltwater, usually forming dense stands similar to Bermuda grass. The plant is of poor to fair grazing value. Blooms summer–fall.

▶ *Sporobolus virginicus* (Virginia dropseed) habit.

▼ *Sporobolus virginicus* (Virginia dropseed) flower.

UNIOLA Sea Oats
Uniola paniculata Linnaeus
⏵ yoo-nee-OH-luh pan-ick-yoo-LAY-tuh

Sea oats is a relatively large perennial grass 30–60 inches (80–150 cm) tall that is probably the best-known grass of Texas beaches, dunes, and shorelines. The plant is strongly rhizomatous with drooping, open, paniculate flowers (inflorescences). Leaf blades are firm, less than ½ inch (1 cm) broad and 32 inches (80 cm) long. The flat spikelets number 10–20 florets or more per spike and are attached to slender branches. Fruits are small in apical clusters and resemble the seed stalk of oats. The plant develops little herbage and is of little forage value. The plants are frequently found on the top of sand dunes along barrier islands and on coastal islands in bays and estuaries. Sea oats is the most important plant in forming barrier island dunes and in their long-term stabilization. Blooms July–August.

◂ *Uniola paniculata* (sea oats) habit on island in the Upper Laguna Madre.

▾ *Uniola paniculata* (sea oats) spikelet and floret (close-up).

TYPHACEAE Cattail Family
TYPHA Cattail
⇒ TY-fuh

T. domingensis Persoon
⇒ doh-ming-EN-sis

T. latifolia Linnaeus
⇒ lat-ih-FOH-lee-uh

Typha is a perennial rhizomatous herb that forms extensive colonies in shallow water. The leaves are two-ranked and linear. Flowers are unisexual, in a dense terminal spike. Staminate flowers are found in the upper portion of the spike, whereas the pistillate flowers are lower on the spike and with spreading bristles that form the down of the fruit.

Key to the Species
1. Male and female spikes separate . . . . . ***T. domingensis***
1. Male and female spikes adjoining . . . . ***T. latifolia***

Typha domingensis (narrow-leaved cattail) and *T. latifolia* (common cattail) are both perennial grasses 6–9 feet (up to 3 m) tall that grow from creeping rhizomes. The sheathing leaves are flat, linear, sessile, and attached in two vertical rows, often exceeding the height of the stem. The inflorescence is a long terminal spike on a jointless stem. The male and female flowers form on the same spike; the male flowers (staminate) are formed on the upper section; the female (pistillate) flowers are located on the lower portion, forming a dark brown spike that varies in size. The difference between the two species is whether or not the male and female spikes are separate or adjoining. Cattails are freshwater plants producing dense patches in shoreline areas where the salinity ranges 0–0.5 ppt (parts per thousand) but may also be found in brackish zones where the salinity may range up to 3.5 ppt. The plant rhizomes are used as a source of food for small mammals and geese. Insect-eating birds forage and nest among stands of cattails.

Typha latifolia (cattail) flower spike and flat leaves.

II | Seagrasses of the Texas Coast

Collecting seagrass and seaweeds in the Upper Laguna Madre.

Introduction to Seagrasses

Seagrasses are flowering plants (angiosperms) that are commonly found submerged in the marine environment. They grow in shallow coastal waters in estuaries, hypersaline lagoons, and brackish-water areas of the Texas coast. They are monocotyledons but not true grasses of the family Poaceae. Seagrass beds rank among some of the most productive ecosystems found in shallow waters and make up one of the most visible and important types of coastal ecosystems. They not only dominate their habitat but will create a physical environment that provides a source of energy from primary productivity on which the entire community depends.

Primary production is the most essential function of the seagrass ecosystem. Data on epiphyte, benthic macrophyte, and microphyte algae and phytoplankton are being investigated. The production of epiphytes can reach 50% of the seagrass production. Moreover, loose-lying algae (drift algae) between the seagrass plants may form a dense layer on the bottom of the seagrass beds, accounting for 10%–20% of the total aboveground biomass. Considering the amount of oxygen produced, the photosynthetic activity of this algal mat is considerable. Data on the productivity of these drift algae are also rare, with research ongoing. Additionally, the productivity of the phytoplankton above and between the seagrass must also be taken into consideration. As an estimate, it is possible that the contribution of the seagrass component to the community productivity may be only 50% of the total productivity in well-structured communities. Epiphytes are extremely affected by nutrient availability and the salinity of the water. The hypersaline conditions of the Upper Laguna Madre have caused lower epiphytic growth on seagrasses than on those in nearby Corpus Christi Bay, which has a lower salinity. The epiphytes found growing on seagrasses are a very important source of energy and are heavily consumed by herbivores.

In order for marine vascular plants to successfully colonize the marine environment, they must be able to (1) tolerate and adapt to a saline medium, (2) grow when completely submerged by water, (3) endure the effects of wind and tidal-produced currents and wave action, and (4) be capable of hydrophilous (submerged) pollination. In addition, the seagrasses must be able to compete successfully in the marine environment.

Worldwide, there are about 50 recognized species of seagrasses that are placed in 12 genera. Seagrasses are not restricted to tropical and subtropical latitudes, but there is a tendency for more species to be found in the tropics. Along the Texas coast five species of seagrasses are found in three families: *Halodule beaudettei* (C. den Hartog) C. den Hartog [= *H. wrightii*; = *Diplanthera wrightii*]; *Cymodocea filiformis* (F. Kützing) D. Correll [= *Syringodium filiformis*]; *Thalassia testudinum* J. Banks & D. Solander ex König; *Halophila engelmannii* P. Ascherson; and *Ruppia maritima* C. Linnaeus.

List of Seagrasses

Family CYMODOCEACEAE
    Cymodocea filiformis (F. Kützing) D. Correll MANATEE GRASS
    Halodule beaudettei (C. den Hartog) C. den Hartog SHOAL GRASS
Family HYDROCHARITACEAE
    Halophila engelmannii P. Ascherson CLOVER GRASS
    Thalassia testudinum J. Banks & D. Solander ex König TURTLE GRASS
Family RUPPIACEAE
    Ruppia maritima C. Linnaeus WIDGEON GRASS

Key to the Seagrasses

1. Leaves terete (cylindrical) . . . . ***C. filiformis***
1. Leaves flat and bladelike or strap-shaped. 2
2. Leaves strap-shaped, less than 1½ inches (4 cm) long, sheath; in whorls or pseudo-whorls; differentiated into a petiole and a blade. . . . . ***Halophila engelmannii***
2. Leaves with a basal sheath enclosing the stem; not arranged in whorls 3
3. Leaves ¼–½ inch (0.5–1 cm) wide; two pericentral veins . . . . ***T. testudinum***
3. Leaf less than ¼ inch (4 mm) wide 4
4. Leaf tip forms three-point crown; two roots per node ***Halodule beaudettei***
4. Leaf tip tapers to a sharp point; one root per node. . . . . ***R. maritima***

Plant Descriptions and Images

Family CYMODOCEACEAE
***CYMODOCEA* Manatee Grass**
Cymodocea filiformis (F. Kützing) D. Correll [*Syringodium filiformis*]
    ➡ sy-moh-DO-see-uh fil-ih-FOR-miss

Manatee grass is easily recognized by its long, 4–12 inches (10–30 cm), terete (cylindrical) leaves with blunt apices that develop in clusters of two or three. The blades, ⅛ inch (1–2 mm) wide, are cylindrical in cross section and have sheaths that completely surround the leaf base. Scale leaves are also present. The rhizome is cylindrical, and growth is a result of damage to the existing meristem or through development of new shoots. Within the terminal (tip) meristem of the rhizome, three roots (at the nodes) are produced. *Cymodocea* flowers occur in unisexual, axillary, cymose clusters on separate plants (dioecious). The female (pistillate) flowers are subtended by hyaline (transparent) bracts and have a short style with two stigmas. The male staminate flower consists of two anthers on the end of a long stalk (pedicel); both are equally attached at the same height. The fruit is about ⅛ inch (3 mm) long and is beaked by a persistent style. Manatee grass is commonly found mixed with other seagrasses in a bed or in small, monospecific patches. The patches often accumulate a large understory of unattached macroalgae. The plants are restricted to shallow areas 2–5 feet (1–2 m) deep with a bottom of mixed sand and mud with salinities ranging about 20–35 ppt (parts per thousand).

Cymodocea filiformis (manatee grass) habit in the Upper Laguna Madre.

Cymodocea filiformis (manatee grass) habit in the Upper Laguna Madre.

Cymodocea filiformis (manatee grass) structure of leaves and stolons.

Cymodocea filiformis (manatee grass) female flowers.

Cymodocea filiformis (manatee grass) male flowers.

HALODULE Shoal Grass
Halodule beaudettei (C. den Hartog) C. den Hartog [*H. wrightii*]
⟹ hal-oh-DOOL-ee bow-DET-ee-eye

Shoal grass is important as an early colonizer of disturbed areas and in locations where *Thalassia* and *Cymodocea* are excluded because of regulating environmental conditions. Therefore, *Halodule* may be considered a pioneering species because it has the ability to establish itself quickly, has coverage rates that are relatively quicker than other competing seagrass species, and will form extensive colonies. *Halodule* is also the most tolerant of all the Texas seagrasses to variations in temperature and salinity. The plant prefers a range of salinity from 20 to well above 40 ppt (parts per thousand) in water 1–3 feet (30–100 cm) deep. In low-salinity areas, care must be taken to avoid confusing it with *Ruppia maritima*.

Shoal grass has narrow, flat, bladelike leaves ⅛ inch (1.5–2 mm) wide and 3–15 inches (7–40 cm) long that are clearly truncate and are three-toothed (one central tooth and two laterals) at the apex. This is an important recognition character formed from the prominent midrib and two smaller parallel marginal veins. This perennial is herbaceous with short, erect stems that emerge from a branching, creeping rhizome that will produce 1–4 leaves. Elliptical scale leaves ¼–½ inch (5–10 mm) long are also present. Flowering has been rarely observed within Texas waters, but when present, these inconspicuous flowers are solitary, located axillary, unisexual with the sexes on two separate plants (dioecious), and are enclosed within a leaf similar to the vegetative leaves. Fruits are globose, ⅛ inch (2 mm) wide, and rarely seen. The primary means of reproduction is vegetative through branched stolons and swollen rhizomes. The plants have creeping roots located at each of the nodes.

Halodule beaudettei (shoal grass) three-point "crown," or blade apex.

Halodule beaudettei (shoal grass) habit.

Halodule beaudettei (shoal grass) habit (close-up).

Family HYDROCHARITACEAE
HALOPHILA Clover Grass or Peanut Grass
Halophila engelmannii P. Ascherson
⟹ hal-oh-FY-luh en-gel-MAN-nee-eye

Clover grass is a dioecious plant with thin, fragile, creeping rhizomes having one root at each node. A terminal rosette of leaves is attached to erect shoots and are pseudo-whorled or in distichously arranged pairs. The leaf blade is oblong or linear-oblong with an obtuse apex and cuneate (triangular) base. The blade margin is finely serrated. Paired scales are found on the stalk at both the base and at midpoint. The female flower consists of a small perianth with a sessile or subsessile, ovoid-shaped, ⅛ inch (3–4 mm) long ovary. The male flower is produced singly on a ¼–½ inch (4–10 mm) long pedicel (stalk). The sepals are broadly elliptic and reflexed when mature. The male anthers are ⅛ inch (4 mm) long, bilocular, and produce yellow pollen grains in fine filaments. *Halophila* must be rooted at the surface oxic zone.

The fruits of *H. engelmannii* are borne on short stalks and resemble the spherical air-filled vesicles of *Sargassum* in size, shape, and ability to float. The fruits are globose to subglobose in shape with diameters between ⅛ inch and ¼ inch (3.0–5.5 mm).

Halophila engelmannii (clover grass) habit.

Halophila engelmannii (clover grass) serrated leaf edge (marginal teeth; close-up).

Halophila engelmannii (clover grass) whorled leaves and stolons.

THALASSIA Turtle Grass
Thalassia testudinum J. Banks & D. Solander ex König
⟹ tha-LASS-ee-uh tes-too-DIN-num

Turtle grass is a favorite food of marine turtles, consequently its common name. This perennial has erect shoots that produce two-ranked clusters of 3–7 broad leaves to ¾ inch (2 cm) wide that develop from a basal meristem. The leaves, up to 14 inches (35 cm) long, have sheaths that enclose the upper portion of the short shoots and arise from a rhizome that is usually buried 1–3 inches (3–15 cm) in the substratum (bottom). The bases of the leaves are usually covered with the remnants of old leaves. The rhizome has apical meristematic tissue that alternates vegetative growth, first branching left, then right. Roots develop from the rhizome near the short shoots. The plant is dioecious with flowers emerging from the base of the leaves. The male flowers have nine stamens with oblong-shaped anthers attached to a pedicel. The pistillate (female) flowers are almost sessile and have 3–5 petals. Flowering is common, occurring when environmental conditions are met. The fruit is beaked and contains 4–5 seeds. Turtle grass will form extensive meadows in coastal areas that have a mud-sand substrate with a water depth of 2–5 feet (60–140 cm) and a salinity range of 20–35 ppt (parts per thousand).

Thalassia testudinum (turtle grass) habit.

Thalassia testudinum (turtle grass) male flower.

Thalassia testudinum (turtle grass) female flower bud.

Thalassia testudinum (turtle grass) flowers: male (left) and female (right).

Thalassia testudinum (turtle grass) fruit (close-up).

Thalassia testudinum (turtle grass) fruit in situ.

Family RUPPIACEAE
RUPPIA **Widgeon Grass**
Ruppia maritima C. Linnaeus
⟹ RUP-pee-a mer-ih-TEE-muh

Widgeon grass is not a true seagrass but rather a freshwater angiosperm that has a pronounced salinity tolerance. It is both eurythermal and euryhaline and capable of growth and completion of its life history in many of the hypersaline bays and estuaries of Texas. *Ruppia maritima* is a perennial with long, branched stems (stolon) up to 3 feet (90 cm), which arise directly from a thin, partially upright rhizome positioned above the substratum (bottom). Roots are one per node. The narrow ⅛ inch (1–1.5 mm) wide pointed leaves are in a cluster of 2–4 that develop directly from a node and are absent of a stalk. Flowering is common, and sexual reproduction appears to play a large part in the propagation of the species. The axillary flowers develop from stems floating on the water's surface. They are small, inconspicuous, and enclosed in semitransparent leaf sheaths. Each cluster of asymmetrical fruits is attached to a terminal stalk. Vegetative reproduction is also a common form of growth in *Ruppia*.

There is sometimes uncertainty in the identification between the seagrasses *Halodule* and *Ruppia*. *Ruppia* is often found adjacent or within *Halodule* beds in regions of reduced salinity. Their external morphology is similar and may cause confusion in identification, but there are five visual clues that separate *Ruppia* from *Halodule*:

1. *Ruppia* produces copious flowering pedicles with numerous seed clusters that may reach 3 feet (1 m) in length. The flowers in *Halodule* are rarely seen.
2. *Ruppia* blades taper to a single sharp point. The blade tip of *Halodule* forms a miniature three-point crown. The two leaf margins and the midvein form the three points.
3. The rhizomes of *Ruppia* are often zigzagged when viewed from above and may be either green or white. *Halodule* rhizomes are usually very straight and white.
4. *Ruppia* has one root per node at the rhizome. *Halodule* has two roots per node.
5. *Ruppia* has extensive aboveground branches and plant material. *Halodule* has only vertical shoots.

Ruppia maritima (widgeon grass) habit.

Ruppia maritima (widgeon grass) branching habit.

Ruppia maritima (widgeon grass) flowers and fruit.

III | Mangroves of the Texas Coast

Red mangrove habitat.

Introduction to Mangroves

Mangrove comes from the two words *mangue,* "tree" (Portuguese for *Rhizophora*), and *grove,* "stand of trees." Mangroves occur in muddy, tidal waters throughout most of the tropical and subtropical world in areas protected from wave action and in zones of high sedimentation. Characteristically, the plants grow in quiet lagoons and/or estuaries and are closely associated with the intertidal marine environment. Many have an intricate growth of prop roots that support the vegetative portion of the plant above a mud bottom or shoreline. The leafy parts of the plant will often hang down to the level of high tide. I have chosen to include information and descriptions of the four common species that can be found in the tropical and subtropical areas of the Gulf of Mexico. It is possible that all four species or their reproductive structures (seeds) may be found in the warmer areas of the Texas coast. All are found in Mexico and Florida's tropical shorelines. As these plants are associated with the marine surroundings, they have adapted to their environment and developed unique mechanisms that enhance survival, including specialized seed dispersal, germination types (including vivipary), specialized salt-regulation mechanisms, and xerophytic (dry) structures.

The major roles of mangrove include substrate formation by trapping debris, filtration of runoff and removal of terrestrial organic matter, creation of habitat and nurseries for other living organisms, and production of detritus that contributes to near and offshore productivity.

List of Mangroves

Family AVICENNIACEAE
Avicennia germinans (Linnaeus) Linnaeus BLACK MANGROVE
Family COMBRETACEAE
Conocarpus erectus Linnaeus BUTTONWOOD
Laguncularia racemosa (Linnaeus) Gaetner WHITE MANGROVE
Family RHIZOPHORACEAE
Rhizophora mangle Linnaeus RED MANGROVE

Key to the Mangroves

1. Leaves alternate, simple, entire, and ovate; fruit persistent woody aggregates . . . .*C. erectus*
1. Leaves opposite; fruit not a persistent woody aggregate 2
2. Leaves whitish on the underside, dark green and lustrous on the upper surface, broadly elliptical; fruit egg-shaped, flattened . . . . **A. germinans**

2. Leaves green on both upper and lower surfaces 3
3. Leaves somewhat lighter green on the underside, ovate; seeds germinating on trees to form an embryo ....*R. mangle*
3. Leaves almost identical on upper and lower surfaces, broadly oblong; fruit about ½ inch (1.5 cm) long with 10 ribs ....*L. racemosa*

Plant Descriptions and Images

Family AVICENNIACEAE
***AVICENNIA* Black Mangrove**
Avicennia germinans (Linnaeus) Linnaeus
⇒ av-ih-SEN-ee-uh JER-min-anz

Black mangrove, the most common mangrove found along the Texas coast, is a maritime shrub or small tree with thick, opposite, entire leaves. The leaves are dark green and shiny (glabrous) on top and gray to whitish below due to a dense covering of club-shaped hairs. The leaves also have specialized salt-secreting structures (hydathodes). The primary stems have a flaky, black bark. The 4–8 perfect flowers are found on a reduced spike in mostly terminal clusters. The corolla is about ½ inch (10–12 mm) wide, white with a yellow throat. The viviparous fruit is a flat, egg-shaped compound capsule that contains a single seed. There are five types of roots: primary adventitious from the stem, cable roots, erect pneumatophores (air roots), descending anchoring roots, and absorbing roots on the pneumatophores. The pneumatophores emerge vertically from horizontal cable roots and may extend 4–12 inches (10–30 cm) above the substrate (and water) and function in gas exchange.

Avicennia germinans (black mangrove) habit.

AVICENNIACEAE

Avicennia germinans (black mangrove) leaves and fruit.

Avicennia germinans (black mangrove) fruit (close-up).

Avicennia germinans (black mangrove) pneumatophores.

Family COMBRETACEAE
CONOCARPUS **Buttonwood**
Conocarpus erectus Linnaeus
⏵ koh-noh-KAR-pus ee-RECK-tus

Buttonwood is found in the upland areas of the shoreline. It is a questionable "true mangrove." The common name comes from the buttonlike appearance of the dense flower heads that grow in branched clusters and form a conelike fruit. The leaves are alternate, an arrangement different from that of the other three species of mangroves, which are opposite. The leathery leaves are simple, 2–3 inches (4–9 cm) long, with a pointed tip, a smooth margin (edge), and wedge-shaped at the base. Two salt glands are located at the base of the blade of each leaf. The small, green flowers are inconspicuous and found in branched terminal panicles (clusters). The fruit is a round and persistent wood aggregate.

Concarpus erectus (buttonwood) flowers.

▼ *Concarpus erectus* (buttonwood) fruit.

LAGUNCULARIA White Mangrove
Laguncularia racemosa (Linnaeus) Gaetner
 ▶ la-gun-koo-LAY-ree-uh ray-see-MO-suh

White mangrove is usually found upland of the black and red mangroves. It has leathery leaves that are opposite, elliptical, and rounded at both ends. A pair of glands on the petioles of the leaves aid in distinguishing this species from other mangroves. Tiny pores on the underside of the leaf excrete salty brine. The leaf also contains tannin cells. The white, bell-shaped flowers are perfect and found in terminal spikes. The 10-ribbed oblong fruit is flattened laterally with two thick and spongy wings. The seeds will germinate (vivipary) while still attached to the tree. Pneumatophores and/or prop roots may be present depending on environmental conditions.

▶ *Laguncularia racemosa* (white mangrove) habit.

▼ *Laguncularia racemosa* (white mangrove) flower.

Family RHIZOPHORACEAE
RHIZOPHORA Red Mangrove
Rhizophora mangle Linnaeus
⟹ ree-ZOH-for-uh MANG-leh

Red mangrove has thick, leathery (coriaceous) leaves that are simple, elliptically shaped, with an entire (smooth), slightly recurved margin (edge) and are arranged opposite, with paired and interlocked stipules. The buds are figlike and located at the apex of each branch. Bisexual flowers are small and develop during the summer months. One or more perfect flowers are formed in the axis of the leaves. Four fleshy, green sepals are fused into a calyx that will persist on the fruit after fertilization and dehiscence (drying) of the four delicate, white petals. Fruit is a single seed that is viviparous; the radicle and hypocotyl may extend out 8 inches (20 cm) from the fruit.

The bark of young roots, stems, and the wood has a red hue and contains large amounts of tannin. Stems lack distinctive growth rings, as the vascular bundles are evenly distributed. The plant has both prop and drop roots with shallow roots that produce anchoring roots.

Rhizophora mangle (red mangrove)
▲◀ flower.
◀ thick and leathery leaf.
▲ fruit (close-up).

RHIZOPHORACEAE

▶ *Rhizophora mangle* (red mangrove) fruit (seeds) germinating on tree.

▼ *Rhizophora mangle* (red mangrove) habit.

Jetty underwater flora.

IV | Common Seaweeds of the Texas Coast

Seaweeds common to the Texas coast.

Introduction to Seaweeds

People have named and classified plants (and animals) for thousands of years. In early systems, taxonomists constructed classification schemes that recognized the similarity of groups and gathered them into clusters of natural assemblages. As science advanced, newer phylogenetic classification schemes have been developed and employed. These modern systems work very well for many organisms; however, the algae (seaweeds) are separated from other groups as a dissimilar group. Newer DNA techniques have brought much information to aid in the classification of the algae but require advanced knowledge of techniques and expensive equipment. Students and individuals who are identifying seaweeds in the field and in a basic laboratory need a method of identification that is reliable yet does not require advanced training and equipment. Along with the keys are images that will provide visual clues to the differences between the various plants. The taxonomic names and arrangement follow Michael J. Wynne's *Checklist of Benthic Marine Algae of the Coast of Texas* (2009). The species covered herein do not include rare or minute species or those found only offshore in the Gulf of Mexico (especially Flower Gardens, a national marine sanctuary). Also, it is important to note that with the exception of a few seaweeds, there are practically no common names used to describe the species, so scientific names are used.

The seaweed (algal) body lacks differentiation into root, stems, and leaves, which separates them from angiosperms (flowering plants). The algae are all photosynthetic plants with chlorophyll *a* and accessory pigments. The organs associated with sexual reproduction and the sporangia are frequently unicellular and, when multicellular, all cells are fertile. The zygotes (fertilized eggs) never develop into multicellular embryos while in the female reproductive organs. This technical definition distinguishes algae from the fungi and angiosperms.

The earliest attempts at subdividing the algae were based on plant color, and three main groups were recognized: red (Rhodophyta), brown (Phaeophyceae), and green (Chlorophyta). The Rhodophyta are the most primitive group of seaweeds, and the Chlorophyta, the most advanced. Present-day identification of the classes is similarly based on differences of color, but using spectrophotometric and chromatographic analysis of the pigments is more accurate than visual examination. The use of these tools has enabled a more precise description of plants. Color differences are due to a predominance of one or another pigment, especially within the brown and red algae. Additional support for separation of the three major groups is based on the different storage reserve products present. A point particularly appropriate to studies on the intertidal algae is that under certain environmental conditions the predominating pigment may be lost, and red or brown seaweeds may become green in color.

The marine macroalgae (seaweeds) are found in all coastal ecosystems, including rocky seashores as well as human-made jetties, salt marshes, seagrass meadows and lagoons, unvegetated bottoms of sand and mud, and, in tropical areas, coral reefs and mangrove communities. In general, the shores of the Texas coast are a soft and sandy or muddy bottom where seaweeds do not grow well. The Texas coast is bordered by barrier islands that essentially are large sandbars with shallow bays and lagoons that separate them from the mainland. A typical beach on the Gulf of Mexico side of the barrier island is very dynamic. The shifting sand and scouring action will usually prevent algal attachment or may bury any algae that successfully attach.

Most seaweed requires some hard or solid surface for attachment. The greatest algal diversity and coverage are found attached to the artificial granite jetties associated with dredged channels of major passes. These provide water movement and tidal exchange from the estuaries out into the Gulf of Mexico, and vice versa (e.g., Galveston, Port Aransas, Port Mansfield, Brazos Santiago Pass). The typical rocky shoreline may be divided into three vertical regions. The intertidal (littoral) zone is exposed during low water and submerged at high water. Along the Texas coast, this vertical distance is usually less than 3 feet (1 m), as the tides are considered to be mixed. Above the intertidal zone is the supralittoral or spray zone. The organisms living here rely on the moisture from the spray of wave action. Below the intertidal zone is the sublittoral zone, where the organisms remain submerged. Note that intertidal boundaries are not fixed elevations, and algae living above the upper intertidal zone may be flooded (covered with water) during spring or storm tides. Algae growing in the lower zones may be exposed to the air only during unusually low tides.

The amount of light available at different water depths is primarily a factor of water turbidity. The result of this turbidity and a short vertical tidal distance results in a compressed zonation of marine plants along much of the Texas coast. Generally, green algae are found in the spray and upper intertidal zones, brown algae are in the submerged zone usually no more than 6 feet (2 m) deep, and the red algae are adapted to all three zones. Some algae have adapted to zones that consistently have high energy by developing a thallus (body) that contains or is covered with calcareous material and may even become encrusting (e.g., coralline red algae, *Padina*).

The soft bottom found in Texas estuaries (bays and lagoons) is usually mud and sand covered with vascular plants, including seagrasses, salt marsh plants, and/or mangroves. They provide submerged surfaces for the growth of epiphytic algae and may trap large amounts of drift macroalgae, especially in areas of low energy. A few algae (e.g., *Caulerpa, Penicillus*) have developed rhizoidal (rootlike) systems that allow them to grow well on sandy bottom in protected areas.

Practically any hard material that is found submerged in the coastal waters may become covered with algae and other organisms. There are also oyster reefs in some bays, serpulid reefs composed of the calcareous external tubes of marine segmented worms (polychaetes) in Baffin Bay, Texas, and shells (scallops, quahogs) that may offer a site of attachment for certain seaweeds (e.g., *Acetabularia, Batophora*).

▲ Jetty along Gulf coast.

▼ Bay mud and shell bottom.

Key to the Divisions of Seaweeds

1. Plants prokaryotic; cells lacking an apparent internal organization; pigments diffused throughout the cytoplasm (not included).
   Division Cyanobacteria........................**Blue-Green Algae**
1. Plants eukaryotic; cells possessing an internal organization with pigments contained within plastids (chloroplasts)..............................2
2. Plants green; plastids more or less grass-green; starch present.
   Division Chlorophyta ....**Green Algae**
2. Plants reddish, brownish, whitish, red-olive; plastids not (usually) grass-green; starch absent...3
3. Chloroplasts brown and plants brown; pigments not soluble in boiling water.
   Division Ochrophyta, Class Phaeophyceae..............**Brown Algae**
3. Chloroplasts red to purple to bluish to red-olive; pigments soluble in boiling water causing plant to become greenish when boiled.
   Division Rhodophyta...................................**Red Algae**

Introduction to the Common Red Seaweeds

The beginning student of phycology (the study of algae and seaweeds) has long been puzzled by the problem of identification of various species of red algae in the field, as many of them are not actually red in color, but green, brown, purple, or even blackish. The typical red color of the group is due to the presence of accessory pigments, including phycoerythrin (red), phycocyanin (blue/violet), and allophycocyanin (blue/violet), which may mask chlorophylls *a* and *d*, the carotenes (orange), and xanthophylls (yellow). So under various environmental conditions, the red pigments may be reduced and the accessory pigments dominate and express various color combinations.

The majority of marine reds have a grooved pit plug between adjacent cells. They are filamentous in construction and usually exhibit apical growth. They range from openly branching filaments to pseudoparenchymatous (mass of filaments in which individual filaments are difficult to distinguish) thalli. The latter may be erect or encrusting and at times covered with calcium carbonate.

The student is further challenged when told that the red algae's primary morphological distinction from other algae is found in the presence of nonmotile male gametes that fuse with a special female sex organ, the carpogonium. These situations often make it difficult for a beginning student to place specimens in the correct algal division. This is especially true for artificial keys designed to key all groups without regard to natural relationships.

Nevertheless, in the use of keys, students will discover numerous reasons to become familiar with some of the basic aspects of reproduction in the red algae. In this group, the reproductive structures are often prominent and provide a suitable character for the recognition of most genera. As a broad generalization, it may be said that a majority of the red algae have a life cycle that is an alternation of not two but three phases (triphasic life history): a sporophyte and a gametophyte generation that are isomorphic (all reproductive cells look alike), and a carposporophyte generation that develops and remains attached to and is dependent on the gametophyte generation. To describe briefly, a macroscopic sporophyte plant produces nonmotile asexual spores, usually tetraspores, which germinate to produce separate male and female gametophyte plants. The male plants produce nonmotile sexual cells (spermatia) that are freed and lodge against and fuse with the female sex organ (carpogonium) on the female plant. As a result of this union of gametes, which may occur in different ways, a new generation begins to develop in or on the female gametophyte. This generation consists of a tissue called a gonimoblast, which produces carpospores in a number of ways. The release of the carpospores and their germination begin the development of the sporophyte generation again, whereby the cycle is repeated.

A review and description of the characters that are used in the following key will provide the basic knowledge necessary for the successful identification of the common red seaweeds found along the Texas coast. Use the Key to the Divisions of Seaweeds to determine that the specimen is a red alga (seaweed). Place the specimen in a pan of shallow water for observation and evaluation. The plant will have one of five body types: (1) filament (slender), (2) internal filaments surrounded within mucilaginous gel, (3) erect bushy mass, (4) membranous (flat sheet), or (5) calcified (small) segments.

Determine if the plant is unbranched or branched. If branched, check to see if the branching is dichotomous, alternate, or irregular. Also find out if there are whorls of spines at the nodes.

In certain red algae, a ring of cells, called pericentral cells, develops around the axial (center) cell. These cells may be visible from the exterior or covered by cortex cells. In some species, there may be as few as 4 or up to 20 pericentral cells per cell. This number is fairly constant in a species and, therefore, is sometimes used as a character to aid in identification. With a hand lens or dissecting microscope, check to see if the filaments are banded with cells at the nodes (corticated). These bands are formed from the orderly division of pericentral cells. To determine the number of pericentral cells, take a short piece of a filament from the alga and place it on a microscope slide. Add a drop of dilute acid (vinegar will do) and let sit for a few seconds. Then, remove most of the liquid with a paper towel and add a cover slip. With the use of the eraser end of a pencil, gently tap the top of the cover slip until you can see the cells separating. You should now be able to count the number of cells. Count the pericentral cells in three or four filament cells and get an average number per cell. This information is used in the key as necessary.

The morphology of reproductive structures is also used to support the correct identification of a red alga. Tetraspores develop from a tetrasporophytic plant, and the four resulting spores are most frequently arranged in one of three different configurations: tetrahedral, zonate, or cruciate (see images). In a few cases, the spores may be single (monospore), in twos (bispores), or in groups larger than four (polyspores). Also, the shape and location of the carposporangium and the morphology of the contained spores may be helpful in the identification of red seaweed.

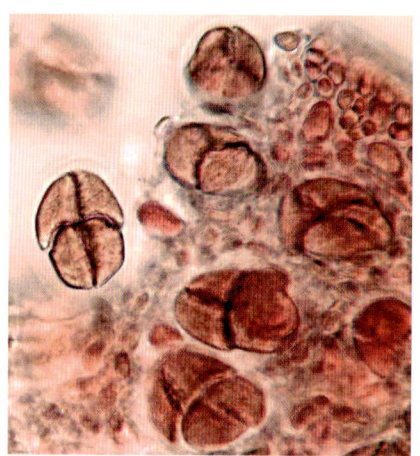

◀ Cruciate tetraspores.

▲ Zonate tetraspores.

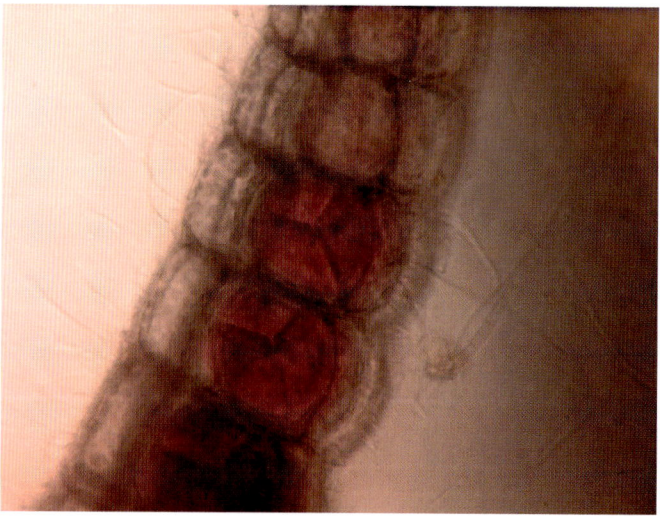

◀ Tetrahedral tetraspores.

List of Common Red Seaweeds

Division Rhodophyta

Class Bangiophyceae

Order Bangiales
Family BANGIACEAE
  Bangia fuscopurpurea (Dillwyn) Lyngbye [*B. atropurpurea*]
  Porphyra rosengurtii Coll et J. Cox [*leucosticta*]
Order Corallinales
Family CORALLINACEAE
  Jania adhaerens J. V. Lamouroux [*ducussato-dichotoma*]
  J. capillacea Harvey [*adhaerens* sensu auct.]
  J. cubensis Mont. ex Kütz. [*Corallina cubensis, Haliptilon cubense*]
  J. subulata J. Ellis et Sol. [*Corallina subulata, Haliptilon subulatum*]
Order Ceramiales
Family CALLITHAMNIACEAE
  Aglaothamnion cordatum (Børgesen) Feldm.-Maz
    [*Callithamnion cordatum*]
  A. halliae (Collins) N. Aponte, D. L. Ballant. & J. N. Norris
    [*Callithamnion byssoides, C. halliae, C. pseudobyssoides*]
Family CERAMIACEAE
  Centroceras clavulatum (C. Agardh in Kunth) Montagne in Durieu
    de Maisonneuve
  Ceramium cimbricum H. E. Petersen in Rosenv. [*C. fastigiatum*]
  C. c. f. *flaccidum* (H. E. Petersen) Furnari et Serio in Cecere et al.
    [*C. fastigiatum* f. *flaccidum*]
  C. deslongchampii Chauv. ex Duby [*C. strictum*]
  C. subtile J. Agardh
Family SPYRIDIACEAE
  Spyridia filamentosa (Wulfen) Harv. in Hook.
  S. hypnoides (Bory in Belanger) Papenfuss [*S. aculeata*]
Family RHODOMELACEAE
  Acanthophora spicifera (Vahl) Børgesen [*A. intermedia*]
  Bryocladia cuspidata (J. Agardh) De Toni
  B. thyrsigera (J. Agardh) F. Schmitz in Falkenb.
  Chondria capillaris (Huds.) M. J. Wynne [*C. tenuissima*]
  C. cincophylla (Melvill) De Toni
  C. dasyphylla (Woodward) C. Agardh
  C. littoralis Harvey

Digenea simplex (Wulfen) C. Agardh
Neosiphonia gorgoniae (Harvey) S. M. Guim. et M. T. Fujii
  [*Polysiphonia gorgoniae*]
N. tepida (Hollenberg) S. M. Guim. et M. T. Fujii [*Polysiphonia tepida, P. howeii*]
Palisada poiteaui (J. V. Lamour.) K. W. Nam [*Chondrophycus poiteaui, Laurencia poiteaui, L. poitei*]
Polysiphonia atlantica Kapraun et J. N. Norris [*macrocarpa*]
P. boldii M. J. Wynne et P. Edwards
P. denudata (Dillwyn) Greville ex Harvey in Hook.
P. echinata Harvey
P. hapalacantha Harvey
P. havanensis Montagne
P. ramentacea Harvey
P. subtilissima Montagne

Order Gelidiales
Family GELIDIACEAE
Gelidium crinale (Turner) Gaillon
G. pusillum (Stackhouse) Le Jolis

Order Gigartinales
Family CYSTOCLONIACEAE (HYPNEACEAE)
Hypnea cornuta (Kützing) J. Agardh
H. musciformis (Wulfen in Jacqu.) J. V. Lamouroux
H. spinella (C. Agardh) Kützing [*H. cervicornis*]
H. valentiae (Turner) Montagne [*H. cornuta*]

Family SOLIERIACEAE
Agardhiella subulata (C. Agardh) Kraft et M. J. Wynne
  [*A. tenera, Solieria tenera*]
Solieria filiformis (Kütz.) P. W. Gabrielson [*Agardhiella tenera*]

Order Gracilariales
Family GRACILARIACEAE
Gracilaria tikvahiae McLachlan [*G. foliifera* var. *angustissima*]
Hydropuntia caudata (J. Agardh) Gurgel et Fredericq
  [*Gracilaria caudata, G. verrucosa*]
H. cornea (J. Agardh) M. J. Wynne [*Gracilaria cornea, G. debilis*]

Order Halymeniales
Family HALYMENIACEAE
Grateloupia filicina (J. V. Lamouroux) C. Agardh
G. pterocladina (M. J. Wynne) S. Kawaguchi et H. W. Wang in Wang et al. [*Prionitis pterocladina*]
Halymenia floridana J. Agardh [*bermudensis*]

Order Rhodymeniales
Family RHODYMENIACEAE
   Botryocladia occidentalis (Børgesen) Kylin
   Rhodymenia pseudopalmata (J. V. Lamouroux) P. C. Silva
Family CHAMPIACEAE
   Champia parvula (C. Agardh) Harvey
Family LOMENTARIACEAE
   Lomentaria baileyana (Harv.) Farl. [*L. uncinata*]

Generic Key to the Common Red Seaweeds

1. Plants consisting of slender filaments.................................2
1. Plants membranous or branched....................................7
2. Plants filamentous, unbranched, dull purple or may be yellow-brown if exposed to sun (1) .... ***Bangia***
2. Plants with branching filaments....................................3
3. Plants composed of slender branching filaments with tips that are branched dichotomously with whorls of spines at the nodes (2) ....
   ....***Centroceras***
3. Plants with a radial organization; erect axes branched, corticated 4
4. Plants mostly irregularly or alternately branched (3) 5
4. Plants mostly branched dichotomously to partially dichotomous.........6
5. Pericentral cells up to 8 (4)............................. ***Polysiphonia***
5. Pericentral cells 8–12 .... ***Neosiphonia***
6. Plants composed of bright rose-red tufts, tufts small, about ¾ inch (1 cm), and much branched, usually dichotomously; main axes and lateral branches more or less uniformly corticated with the cortication in most cases restricted to the nodes (4) .... ***Ceramium***
6. Plants composed of rose-red tufts up to 2 inches (5 cm) in height, branching alternate to partially dichotomous .... ***Aglaothamnion***
7. Plants flat, branched or unbranched (1) 8
7. Plants bushy-branched, either calcified or noncalcified..................9
8. Plants a dull pink to dark reddish-purple; oblong blades soft-slippery and very thin; pit connections absent (7) .... ***Porphyra***
8. Plants bright reddish-purple; blades about ¼ inch (0.5 cm) wide, branched dichotomously, particularly near the tips .... ***Rhodymenia***
9. Plants branched with small, highly calcified segments (7)........... ***Jania***
9. Plants composed of noncalcified branches, flattened or cylindrical 10
10. Plants composed of somewhat slender, flattened branches, small to 2½ inches (6 cm) tall; wiry, dark purple to brownish-black, branches erect and flat; rhizines mostly in outer medulla (9).... ***Gelidium***

10. Plants composed of cylindrical axes and branches 11
11. Plants very bushy, some with definitely recurved tips, others with stellate branchlets, often forming an entangled mass, erect branches 4–20 inches (10–50 cm) tall; tetrasporangia zonate (10) . . . . *Hypnea*
11. Plants forming an erect, bushy mass 12
12. Plants forming an erect, bushy mass (11) 13
12. Plant thallus or blade with sparse internal filaments within mucilaginous gel 15
13. Erect branches have short pinnate branchlets; main axes bearing numerous short more or less determinate axes (12). . . *Bryocladia*
13. Plants not as described above 14
14. Plants erect and bushy, dull reddish-brown, wiry below and cartilaginous above; branches covered with slender, short, stiff branchlets; pericentral cells visible only in the shorter lateral branches (13) . . . . *Digenea*
14. Plant rose-red, tends to break easily when handled; branching is radial and corticated . . . . *Spyridia*
15. Thallus of wide, thin, broadly lobed blades 8 inches (20 cm) high (12) . . . . *Halymenia*
15. Thallus with branches 16
16. Plants small, less than 3 inches (8 cm) in height (15) 17
16. Plants taller than 3 inches (8 cm) 18
17. Plants regularly constricted into series of more or less moniliform segments by regular internal septation (16) . . . . *Champia*
17. Plants not constricted into moniliform segments; septae mostly at the base of branches . . . . *Lomentaria*
18. Plants repeated-branched, hollow, and with numerous short, bulbous branches (16) . . . . *Botryocladia*
18. Plants not hollow but containing septa (cross-walls) 19
19. Plants branched, slender to coarse, terete to strap-shaped, prominent recurved branches (18) 20
19. Plants with dense lateral branches 24
20. Plants with strap-shaped branches, firmly gelatinous or fleshy, branching usually pinnate; color dark purple, reddish, or greenish; sporangia cruciate; cystocarps immersed in the thallus (19) . . . . *Grateloupia*
20. Plants deep rose, sometimes translucent; branches constricted at the base, never bearing spines; cystocarps immersed in thallus 21
21. Thallus terete throughout with tapering branches . . . . *Hydropuntia*
21. Plant often with compressed (flattened) main axis 22
22. Plant flattened or at least the portion immediately below a fork (21) . . . . *Gracilaria*

22. Plants with cylindrical radial branches 23
23. Holdfast a complex system of many secondary rhizoids (22)...... *Solieria*
23. Holdfast simple ....*Agardhiella*
24. Lateral branchlets not densely covering the main axes, lateral branches short, blunt, and peglike; plants in cross section without a distinct central axis and without pericentral cells (19) ....*Palisada*
24. Lateral branchlets intergrading in size gradually with main axis........25
25. Lateral branchlets cylindrical and constricted at the base, tips of branchlets blunt (24) ....*Chondria*
25. Lateral branchlets not in rows, sometimes radial containing spurlike spines .... *Acanthophora*

Plant Descriptions and Images

Division Rhodophyta

Class Bangiophyceae

Order Bangiales
Family BANGIACEAE
BANGIA
Bangia fuscopurpurea (Dillwyn) Lyngbye [*B. atropurpurea*]
⟹ BANG-ee-uh fus-co-pur-PURR-ee-uh

This plant consists of unbranched, reddish-purple, fine filaments that are soft, slippery, and may reach a length of about 1 inch (3 cm). The alga is attached by a holdfast of basal rhizoids. The young filaments are uniseriate but with maturity become multiseriate (multicellular) and often knobby. The uninucleate cells contain a single stellate plastid with a pyrenoid. Two types of spores are found. Monospores are produced by the direct transformation of surface vegetative cells or sexually by a shell-boring microscopic phase (Conchocelis phase). The species forms blackish-purple mats on the rocks in the upper intertidal zone on jetties during the winter and late spring.

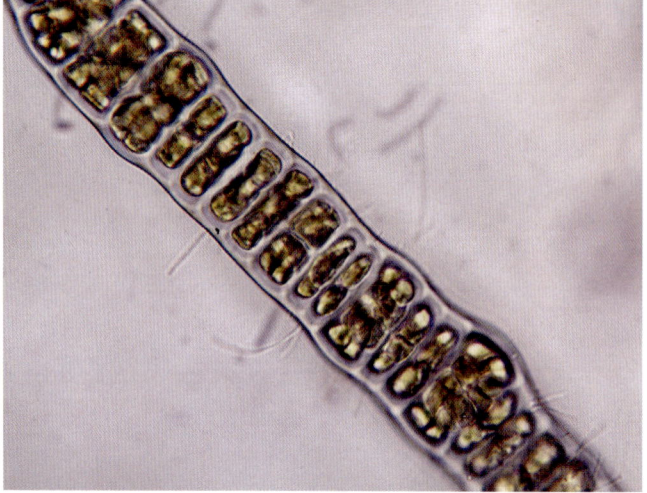

Bangia fuscopurpurea filament developing into multiseriate filament.

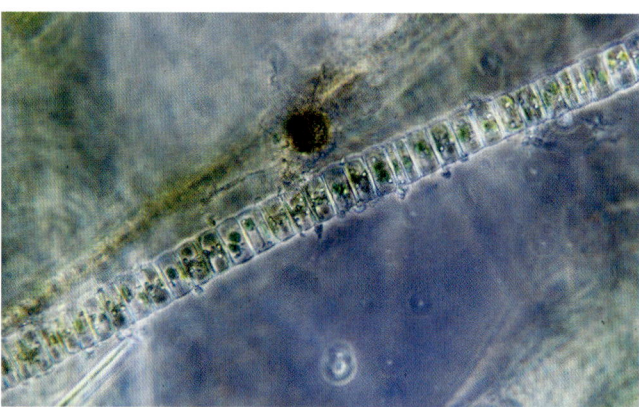

Bangia fuscopurpurea low power (100×).

PORPHYRA
Porphyra rosengurtii Coll et J. Cox [*leucosticta*]
⏵ por-FY-ra roh-SEN-gurt-ee-eye

The plant has a delicate pink to reddish-purple monostromatic (one cell layer thick) membranous blade up to 6 inches (15 cm) long and 5 inches (13 cm) wide. Several blades are attached by a common basal holdfast. The shape of the blade varies from oblong to nearly round. The cells have a single stellate chloroplast with a central pyrenoid. The spermatangia are found in elongated patches, parallel to each other and the blade axis. These algae are found growing attached to the jetties in the upper intertidal zone and form a dense, slippery mat. This is a cold-season plant and is most often collected in late winter.

Porphyra rosengurtii (pressed specimen).

Order Corallinales
Family CORALLINACEAE
JANIA
▶ jan-EE-uh
J. adhaerens J. V. Lamouroux [*ducussato-dichotoma*]
▶ ad-heer-ENZ
J. capillacea Harvey [*adhaerens* sensu auct.]
▶ kap-il-LAY-see-uh
J. cubensis Mont. ex Kütz. [*Corallina cubensis, Haliptilon cubense*]
▶ koo-BEN-sis
J. subulata J. Ellis et Sol. [*Corallina subulata, Haliptilon subulatum*]
▶ sub-yoo-LAH-tuh

Key to the Species
1. Branching irregular .... *J. cubensis*
1. Branching regularly dichotomous 2
2. Segments flattened .... *J. subulata*
2. Segments mostly terete 3
3. Not delicate, to 1½ inches (4 cm) high; segments greater than 100 µm in diameter .... *J. adhaerens*
3. Delicate to ¼ inch (1 cm) high, segments to 100 µm in diameter, form spheres .... *J. capillacea*

J. adhaerens. An erect, brittle, coralline alga with purple-red to reddish-pink to chalky-white calcium carbonate thalli that are dichotomously branched at a wide angle. The plant segments are calcified except at the nodes (geniculate). The segments are 4–6 diameters long. Tetrasporangia are egg-shaped with zonately divided spores. The plant reaches a height of about 1½ inches (4 cm) and occurs as an epiphyte upon seagrasses and other algal species.

J. capillacea. A small, delicate plant with terete axes. The calcified, dichotomously branched thallus (at wide angles) forms tightly clumped, free-floating spheres. The branches are made up of segments 4–6 diameters long and are often recurved with pointed to occasionally rounded apices. The tetrasporangial conceptacles are solitary, inflated segments (sometimes vase-shaped) with a central pore and form zonate tetraspores. Often the conceptacles will bear two hornlike projections that may develop into branches. The plants are found in Texas bays and the Laguna Madre either as an epiphyte on seagrasses and other algae or as free-floating spheres as an important component of the drift algal community. In the shallow water, the plants are overexposed to light, lose their pigments, and turn a chalky-white color. The plant is found throughout the year but reaches it maximum development in late spring.

J. cubensis. An erect plant with creeping, flabellate, delicate, pinkish to rosy-red tufts. Basal crusts give rise to several crowded erect axes that are loosely pinnate or irregularly branched, typically in one plane. The plant segments are calcified except at the nodes (geniculate). The main axes are terete to subterete with obtuse apices. Zonate tetrasporangia are formed in urn-shaped, single-pored conceptacles. The species is a common inhabitant of Gulf of Mexico jetties in the lower zone of algal growth or may be found in bays attached to concrete pilings and walls. The alga is found throughout the year but reaches maximum development during the summer. In bays, the alga occurs from late spring throughout the summer.

J. subulata. A seaweed that forms pinkish-purple tufts up to 1½ inches (3.5 cm) long. The plant is regularly dichotomously branched with pinnately arranged branchlets. The axes are distinctly flattened with segments that are calcified except at the nodes (geniculate). Conceptacles are swollen, terminate the branchlets, and often have one or two hornlike projections. The alga is present throughout the year in the lower zone of algal growth on Gulf of Mexico jetties.

NOTE: *Jania rubens* has been reported from the deepwater reefs of the Flower Gardens.

Jania subulata calcium carbonate branches (close-up; pressed specimen).

CORALLINACEAE

Jania subulata branching habit (pressed specimen).

Jania capillacea habit as drift algae in seagrass bed.

▶ *Jania capillacea* branching as shore debris.

▼ *Jania cubensis* branching in seagrass bed.

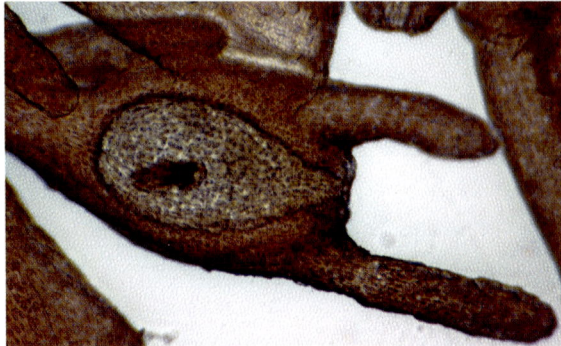

Jania capillacea carposporangium.

Jania adhaerens branching (geniculate).

Order Ceramiales
Family CALLITHAMNIACEAE
AGLAOTHAMNION
⟹ ag-lay-oh-THAM-nee-on
A. cordatum (Børgesen) Feldm.-Maz [*Callithamnion cordatum*]
⟹ kor-DAY-tum
A. halliae (Collins) N. Aponte, D. L. Ballant. & J. N. Norris [*Callithamnion byssoides, C. halliae, C. pseudobyssoides*]
⟹ HALL-ee-ay

Key to the Species
1. Branching pseudodichotomous near the apices; ultimate cells less than 10 µm in diameter; branchlets with incurved tips and not in flat-topped clusters.................................***A. cordatum***
1. Branching alternate near the apices, ultimate cells greater than 10 µm in diameter; branchlets bluntly pointed ....***A. halliae***

These algae form small, delicate, pinkish-red tufts up to about 1 inch (3 cm) in height. The tufts are composed of delicate, pinnately branched filaments. Each cell of the main axis bears a branchlet; each successive branchlet is arranged alternately and has curved tips. Cortication is not present. Reproductive structures include paired carpospores found laterally along the main axes and tetrahedrally divided tetraspores formed on the upper surface of the branchlets. The plant occurs on hard substrates or as an epiphyte on larger algae in the intertidal zone of the jetties. In the bays, this alga is attached to rocks, pilings, and shells and usually becomes much better developed than it does on a jetty. This alga reaches its maximum development in late spring and summer but can usually be found throughout the year.

Aglaothamnion cordatum epiphytic on a red alga.

Aglaothamnion cordatum branching.

Family CERAMIACEAE
CENTROCERAS
Centroceras clavulatum (C. Agardh in Kunth) Montagne in Durieu de Maisonneuve

⏩ sen-troh-SER-as KLAV-yoo-lay-tum

The plant consists of reddish to purple tufts up to 3 inches (8 cm) long. The filaments are narrowly dichotomously branched with forked (pincerlike) apices. The uniseriate axes are entirely corticated by rectangular cells with nodes bearing whorls of one- or two-celled short, colorless spines. The presence of spines at the nodes separates this plant from similar members of the genus *Ceramium*. The filaments are thin and brittle, tend to fragment, and are lost from the plant when it is handled. Tetrasporangia are tetrahedrally divided cells found as whorls on the outer joints. Spermatangia are in terminal clusters. The alga occurs on jetties in the upper zone of algal growth. It is also found growing attached to rocks, shells, pilings, seagrasses, and other larger seaweeds in bays. This alga is found throughout the year but reaches its maximum growth and coverage during the warmer seasons.

▲ *Centroceras clavulatum* branching habit (pressed specimen).

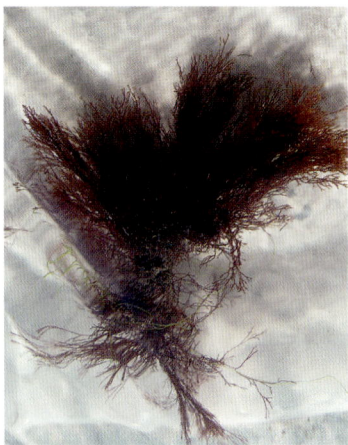

◀ *Centroceras clavulatum* in situ.

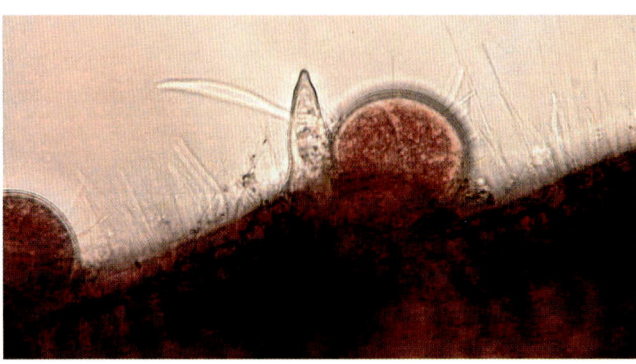

◀ *Centroceras clavulatum* spine at node with tetrahedral tetraspore.

CERAMIUM
⇒ ser-AY-mee-um

C. cimbricum H. E. Petersen in Rosenv. [*C. fastigiatum*]
⇒ sim-BRIK-um

C. c. f. flaccidum (H. E. Petersen) Furnari et Serio in Cecere et al. [*C. fastigiatum* f. *flaccidum*]
⇒ [for-ma] FLA-sih-dum

C. deslongchampii Chauv. ex Duby [*C. strictum*]
⇒ des-LONG-cham-pee-ee-eye

C. subtile J. Agardh
⇒ sub-TIH-lee

The plants of this genus are composed of reddish tufts up to 1 inch (2.5 cm) long of irregular to dichotomous to pseudodichotomously branched, uniseriate filaments that typically have distinctive, pincerlike apices. The cell nodes are corticated with zone(s) of smaller cells, whose number of rows, arrangement, and cell size is important to determine the species within the genus. Tetrasporangia are spherical and tetrahedrally divided. The plants occur on rocks and epiphytically on larger algae in the upper zone on jetties. Maximum development takes place during the warmer parts of the year.

Key to the Species
1. Cortical band consisting of two transverse rows of cells . . . . . **C. cimbricum**
1. Cortical band consisting of more than two transverse rows of cells 2
2. Segments greater than 185 μm in diameter with lower row of transverse cortical cells not broader than long . . . . **C. deslongchampii**
2. Segments greater than 185 μm in diameter 3
3. Lower row of transverse cortical cells distinctly broader than long; branching irregular with slightly incurved apices, rarely forming pincers . . . . **C. cimbricum** f. *flaccidum*
3. Branching dichotomous with small cortical cells at joints; apices strongly incurved, forming pincers . . . . **C. subtile**

C. cimbricum. The nodal cells are arranged in two transverse rows in which the cells in the upper row are usually smaller than those in the lower. The tetrasporangia strongly project from the nodes.

C. c. f. flaccidum. The corticating cells are arranged in several (3–6) rows with a series of transversely elongated cells occurring in the lower portion of each band. The branching is irregular to pseudodichotomous. The apices are only somewhat incurved and rarely form pincers. One or two tetrasporangia are found at the nodes and contain tetrahedrally divided tetraspores.

C. deslongchampii. The nodes are composed of several rows of cells and are readily visible to the naked eye. The tetrasporangia are immersed at the nodes and arranged in a single whorl per segment.

C. subtile. The thallus is bushy, forming tufts to 6 inches (15 cm) high. The widely dichotomously branched filaments are a dull brownish-red. The branches have small corticating cells at the joints with pincers with strongly incurved apices. The tetrasporangia are spherical and found solitary protruding from joints.

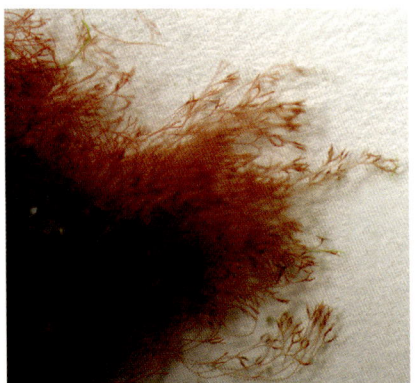

Ceramium cimbricum branching habit.

Ceramium cimbricum cortication with cruciate tetraspores.

▲ *Ceramium cimbricum* cortication.

▶

Ceramium cimbricum cruciate tetraspores.

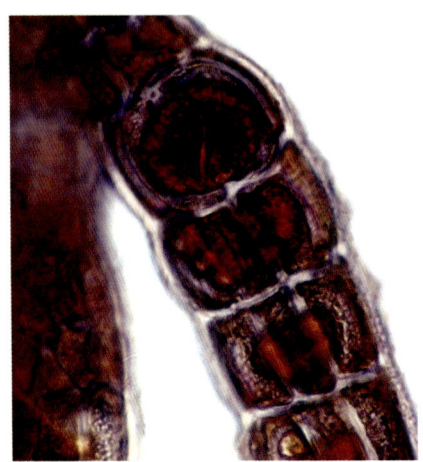

Family SPYRIDIACEAE
SPYRIDIA
⇒ spy-RID-ee-ah

S. filamentosa (Wulfen) Harv. in Hook.
⇒ fil-uh-men-TOH-suh

S. hypnoides (Bory in Belanger) Papenfuss [*S. aculeata*]
⇒ hip-NOY-deez

These large, uniseriate, filamentous plants are rose-red to purple and consist of branches composed of large cells entirely covered with a small-celled cortex of longitudinally elongated cells. Branches are alternately arranged from a distinct main axis. The primary branches are covered with radially positioned, fine, determinate branchlets composed of a single row of cells with one to two rows of cortical cells that encircle the nodes. The branchlets are terminated by a spine. The plant is found attached to jetty rocks and nearby pilings in the lower zone of algal growth. Although found during most of the year, it reaches its maximum development in the fall and early winter. Plants growing in shallower waters tend to be bushier than plants found in deeper water.

Key to the Species
1. Plants without pronounced main axes; branching tips curved, often with distinct hooks. . . . .*S. filamentosa*
1. Plants with pronounced main axes and without curved tips . . . **S. hypnoides**

S. filamentosa. A fuzzy alga up to 12 inches (30 cm) long and dull brown to pale pink. The branching is alternate and may become tangled. The species does not have pronounced main axes and may collapse when removed from the water. Tetrasporangia are tetrahedrally divided and found at the joint of branchlets. The plant is common and present in bays during the warmer months.

S. hypnoides. A densely bushy plant up to 2–4 inches (5–10 cm) long and pink to dark red. The alga has distinct main axes and is more vigorous than the other species. Recurved spines or hooks may be present on the apical cell. Caution: Do not confuse this plant with *Hypnea,* which also has incurved apices (hooks). The alga is found in the upper intertidal zone on Gulf of Mexico jetties during the summer and fall.

Spyridia hypnoides branching (pressed specimen).

Spyridia hypnoides determinant branching.

Spyridia filamentosa branchlets in situ.

Family RHODOMELACEAE
ACANTHOPHORA
Acanthophora spicifera (Vahl) Børgesen [*A. intermedia*]
⟹ a-kanth-oh-FOR-uh spik-EE-fer-uh

Plants are erect, spiny, and coarse with variable coloration ranging from reddish-brown, green to yellow, or occasionally translucent. The branching is radial but at times irregular with terminal branches covered with short, spiny branchlets. The polysiphonous central cells are surrounded with heavy cortication. Tetrasporangia are tetrahedrally divided in short, spiny, swollen branchlets. The cystocarps are urn-shaped and found in the axes of spiny branchlets. Plants may reach a height of 10 inches (25 cm) and have a disklike holdfast. The plants are found attached to hard substrate in calm waters or as drift algae in seagrass beds located near channels that open to the Gulf of Mexico. The plant is common in the Lower Laguna Madre, especially during the summer, and only recently (2008) has been recorded in the Upper Laguna Madre near the new jetties at Packery Channel.

Acanthophora spicifera spiked branchlets (close-up).

BRYOCLADIA
⇒ bry-oh-KLAD-ee-uh

B. cuspidata (J. Agardh) De Toni
⇒ kus-pi-DA-tuh

B. thyrsigera (J. Agardh) F. Schmitz in Falkenb.
⇒ thur-SEE-ger-uh

Key to the Species
1. Axes dense with short, straight to recurved, polysiphonous branchlets; 6–8 pericentral cells . . . . *B. cuspidata*
1. Axes sparse with relaxed, pinnately branched, polysiphonous branchlets; 9–12 pericentral cells. . . . . *B. thyrsigera*

B. cuspidata. The plant has uniformly short branchlets and is the more common of the two species. Both are jetty species found in the middle to lower zone of algal growth. *Bryocladia cuspidata* either appears as tufts or may form extensive mats on the granite rocks. Growth occurs throughout the year but is more abundant in the late summer to fall.

B. thyrsigera. When mature, plants are 3–4 inches (8–10 cm) tall with brownish-black, stiff, cylindrical erect branches. The branches are not corticated but are covered with short branchlets giving the appearance of bristles. Colorless, branched trichoblasts are often present with some near the apices modified into antheridia. Large, urn-shaped cystocarps develop on some of the branchlets. The tetrasporangia are arranged in the upper segments of the branchlets and are divided tetrahedrally.

Bryocladia cuspidata dense branching habit.

Bryocladia cuspidata shoreline habit.

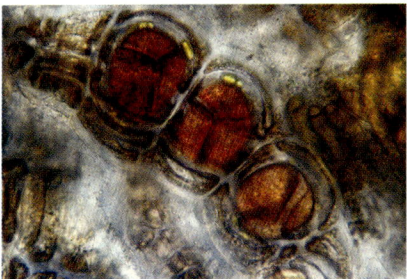

▼ *Bryocladia cuspidata* cruciate tetraspores.

▲ *Bryocladia thyrsigera* carposporangium with carpospores.

CHONDRIA
⟹ KON-dree-ah

C. capillaris (Huds.) M. J. Wynne [*C. tenuissima*]
⟹ kap-ill-AIR-iss

C. cincophylla (Melvill) De Toni
⟹ sing-KOH-fil-uh

C. dasyphylla (Woodward) C. Agardh
⟹ das-ee-FIL-uh

C. littoralis Harvey
⟹ lit-tor-AY-liss

These plants consist of terete, branched axes that bear lateral, spindle- or club-shaped branchlets with the apical cell either exposed on acute tips or sunken in pits of obtuse tips. Both are surrounded by terminal tufts of fine filaments (trichoblasts). A central axial filament is surrounded by several layers of generally smaller cells and ultimately by a cortex of still smaller, pigmented cells.

Key to the Species
1. Apices obtuse or truncate, apical cell sunken in terminal pit 2
1. Apices acute to acuminate or cuspidate, apical cell exposed,
    occasionally hidden by numerous trichoblasts 3
2. Branches 0.50 to 1.25 mm in diameter; new branchlets arising from
    the base of older branchlets; pericentral cells 6–8 (1) . . . . . . ***C. cincophylla***
2. Branches 1.0–1.5 mm in diameter; older branchlets not propagating
    new branchlets from branchlet base; pericentral cells 4–6 . . . ***C. dasyphylla***
3. Plants with branches similar in size to the axes that bear them (1) . . . .
    . . . . ***C. capillaris***
3. Plants with branches markedly more slender than the axes that
    bear them and tapered at both ends . . . . ***C. littoralis***

C. capillaris. A plant with a thallus that is sparse to bushy, translucent yellow to a dull purple, up to 10 inches (25 cm) long. The branches are similar in size to the main axis with cylindrical branchlets that are numerous and clearly constricted at the base and tapered toward a blunt apex. The apical cell is exposed (not in a pit) but may be obscured by numerous dichotomously branched apical filaments. The holdfast is disklike. Tetrasporangia are scattered in the distal ends of the outer branches and tetrahedrally divided. Cystocarps are oval to urn-shaped and scattered in the outer branches or the main axis. The plants are usually found in the intertidal zone and attached to hard substrates.

C. cincophylla. An alga up to 8 inches (20 cm) long, brownish to translucent yellow, with red apices, and has a sprawling, tangled arrangement. The cylindrical branches are alternately to irregularly branched and have 6–8 pericentral cells that are difficult to distinguish. The branchlets decrease in size gradually with the main axes and are constricted at the base, with blunt apices. The terminal apical cells are sunken in an apical pit surrounded by numerous dichotomously branched filaments. The plant does not strongly stain paper brown upon drying. The tetrahedrally divided tetrasporangia are found on fertile branchlets. The plant is found drifting in bays from November to June.

C. dasyphylla. An alga up to 2 inches (5 cm) long and light yellow-brown to red with irregular branching. The plant has a triangular outline with a regular decrease in size between the main axes and the branchlets. The branches are cylindrical with 4–6 pericentral cells. The branchlets are club-shaped, constricted at the base, with truncate tips. The apical cell is located in a terminal apical depression surrounded by numerous, fine dichotomously branched filaments. Tetrasporangia are scattered in the distal ends of the outer branches and tetrahedrally divided. The species stains the paper brown when pressed. The plants are typically collected from the lower intertidal zone on area jetties.

C. littoralis. An alga up to 8 inches (20 cm) high, a light brownish-yellow to brownish-purple with alternate branching. The branches have 5–6 pericentral cells. The species may be readily recognized, as the branchlets are obviously more slender than the main axes and are located only on the outside branches. The branchlets are tapered at both ends, constricted at the base, and have pointed apices with prominent terminal apical cells that are exposed. Tetraspores are tetrahedrally divided. The carpospores are solitary, stalked, spherical to urn-shaped, and found on outer branchlets. The species is found attached to oysters, rocks, and wooden pilings in bays throughout the year.

RHODOMELACEAE

Chondria littoralis in situ.

Chondria littoralis (close-up).

▲ *Chondria littoralis* tetrahedral tetraspores.
◄▲ *Chondria littoralis* branchlet.

◄ *Chondria littoralis* branchlet constricted at base.

▼ *Chondria dasyphylla* tetraspores and apical depression with branched filaments.

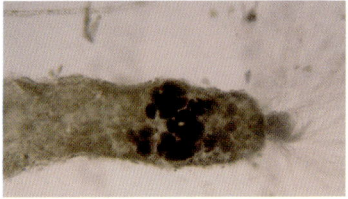

DIGENEA
Digenea simplex (Wulfen) C. Agardh
⟹ dy-GEN-ee-a SIM-pleks

This is a large, 1–10 inches (3–25 cm), coarse alga that is wiry below and cartilaginous above. It is brownish-red to red. The upper portions of the axis and main branches are covered with slender, stiff branchlets. The branchlets are numerous and arranged radially. The pericentral cells are elongated. Antheridia are produced at the apical end of the branchlets. Cystocarps are also found on these branchlets. The tetraspores are tetrahedral in morphology and will develop in the distal portion of the branchlets. The plants occur in bays and estuaries and are usually attached to shells. As the plant grows in size, the shell may break loose from the substrate or the plant will detach from the shell and become part of the drift algal community found in seagrass beds. The alga is common throughout bays and the Laguna Madre during most of the year but reaches maximum growth during the summer and early fall. It is also a host for attachment by smaller algae and invertebrates. The plant is thought to be an indicator of poor water quality and/or higher-than-normal nutrient levels in the water.

Digenea simplex habit.

Digenea simplex branching and branchlets (pressed specimen).

Digenea simplex (close-up).

NEOSIPHONIA
⇒ nee-oh-sy-FON-ee-a

N. gorgoniae (Harvey) S. M. Guim. et M. T. Fujii [*Polysiphonia gorgoniae*]
⇒ gor-GOH-nee-uh

N. tepida (Hollenberg) S. M. Guim. et M. T. Fujii [*Polysiphonia tepida, P. howeii*]
⇒ te-PID-uh

Key to the Species
1. Plants up to 1 inch (3 cm) long; epiphytic on *Halodule beaudettei* . . . . ***N. gorgoniae***
1. Plants usually much longer than 1 inch (3 cm); not epiphytic on *H. beaudettei* . . . . ***N. tepida***

N. gorgoniae. The plants are light brown and up to 1 inch (3 cm) long. Prostrate axes are poorly developed or absent. The polysiphonous branches arise in the axils of the trichoblasts. A few short, spinelike branches are generally present. The main axes are up to 250 μ in diameter, and the segments have a maximum length of 1–2 diameters. Four pericentral cells are present. The antheridial branches occur singly with trichoblasts. The alga occurs epiphytically on the seagrass *H. beaudettei* in bays during the warmer months of the year.

N. tepida. The plant is bright red and up to ¾ inch (2 cm) long. The branches arise in the axils of the trichoblasts. The main axes are up to 100 μ in diameter, and the segments have a maximum length of 2–3 diameters. Eight pericentral cells (occasionally seven) are present. The species is observed in the upper zones of coastal jetties and in area bays during the warmer part of the year.

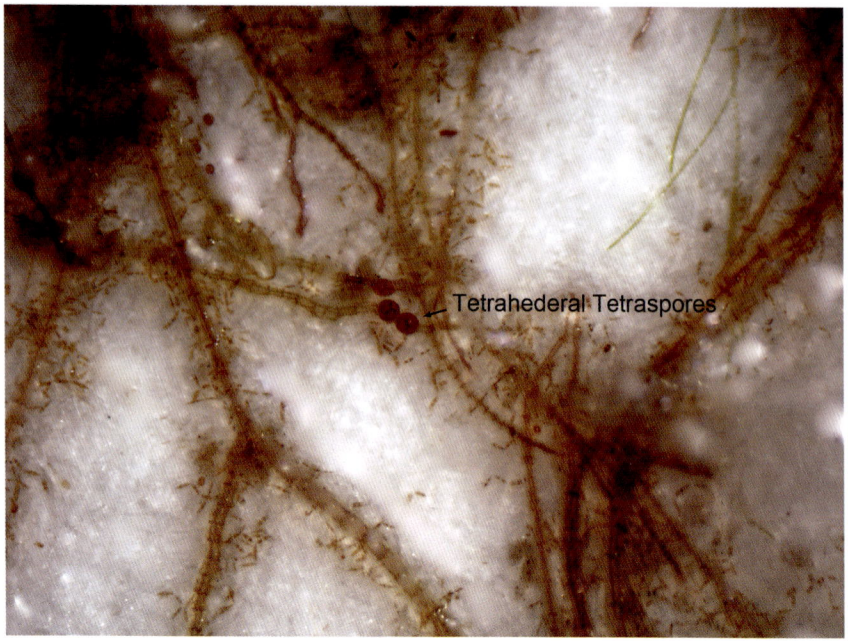

Neosiphonia gorgoniae branching with tetraspores.

Neosiphonia gorgoniae plant as epiphyte on seagrass.

PALISADA

Palisada poiteaui (J. V. Lamour.) K. W. Nam [*Chondrophycus poiteaui, Laurencia poiteaui, L. poitei*]

⇒ pa-lee-SAH-da poy-tee-ee-eye

This is an erect, bushy, cartilaginous plant of variable color that may range from golden-brown to green to brownish to dark purple, often encrusted with calcareous red algae or bryozoans. Plants may reach a length of 6–8 inches (15–20 cm). The axis of the plant is cylindrical and irregularly branched. A central axial filament is surrounded by pericentral cells in the young plant, but the adult axis consists of a medulla of large, colorless cells covered by a cortex of smaller pigmented cells. These branches bear short, lateral, peglike branchlets that are not constricted at the base (as in *Chondria*) and have an apical cell sunk in the depression surrounded by trichoblasts. The antheridia are found on branched trichoblasts in the apical depressions of fertile branchlets. Cystocarps are prominent and scattered over the lateral branchlets. The tetrahedrally divided tetrasporangia are scattered among the cortical cells of end branchlets. *Palisada poiteaui* is a common inhabitant of the Laguna Madre (and other bays), where it is a dominant drift alga found along the edges or within sandy patches and depressions of seagrass beds. Common throughout the year, it reaches maximum growth during the summer.

Palisada poiteaui branchlets (close-up).

◀ *Palisada poiteaui* branching habit (pressed specimen).

▼ *Palisada poiteaui* in situ in manatee seagrass bed.

▼ *Palisada poiteaui* variable pigmentation and branching in *Halodule* seagrass bed.

POLYSIPHONIA
⮕ pol-ee-sy-FON-ee-a
P. atlantica Kapraun et J. N. Norris [*macrocarpa*]
⮕ at-LAN-tik-uh
P. boldii M. J. Wynne et P. Edwards
⮕ BOLD-ee-eye
P. denudata (Dillwyn) Greville ex Harvey in Hook.
⮕ dee-noo-DAY-ta
P. echinata Harvey
⮕ ek-in-AY-tuh
P. hapalacantha Harvey
⮕ hap-ay-luh-KANTH-uh
P. havanensis Montagne
⮕ hav-a-NEN-sis
P. ramentacea Harvey
⮕ ra-men-TAY-see-a
P. subtilissima Montagne
⮕ sub-til-ISS-ih-muh

This genus of algae is easily recognized; however, the identification to species is somewhat difficult and requires the use of a microscope and specialized techniques. Even with proper equipment, the identifications are complicated and may be imperfect because the characters of identification are variable. The bright red to reddish-brown plants are small and slender. The branches are erect, filamentous, and cylindrical. The species can be found throughout the Laguna Madre and bay systems and attached to the jetties. On the jetties, it is most often found in the mid to lower zones of algal growth. Typically, *P. denudata* and *P. atlantica* are collected on the jetties and *P. subtilissima* is collected in the bays and Laguna Madre. The plants are found throughout the year. They are also often epiphytes of other algae. Identification to species requires that the number of pericentral cells per axial segment be known, in addition to other characters. The procedure is to place a section of the thallus on a microscope slide, add one drop of dilute acid, and place a cover slip over the specimen. Press with the eraser end of a pencil to spread out the pericentral cells. Count the number of pericentral cells in four or five cells and determine the average number. Go to the key and try!

Key to the Species
1. Pericentral cells four 2
1. Pericentral cells more than four, usually six 7

2. Main axes covered with short, spinelike branchlets about 1 mm in length (1) 3
2. Main axes not covered with short, spinelike branches. 6
3. Main axes corticated (2) 4
3. Main axes uncorticated; branching somewhat dichotomous ....
    .... *P. atlantica*
4. Main branches strongly corticated and with bristlelike branchlets (3) ......
    .... *P. ramentacea*
4. Main branches lightly corticated 5
5. Main axes covered with short, spinelike branchlets about 1 mm in length (4) .... *P. echinata*
5. Plants delicate, with many branchlets about 2 mm in length ....
    .... *P. hapalacantha*
6. Branches arising in axils of trichoblasts (2) .... *P. havanensis*
6. Branches arising independently of trichoblasts. .... *P. subtilissima*
7. Branches arising in axils of trichoblasts; base a discoid holdfast (1) ....
    .... *P. denudata*
7. Branches arising independently of trichoblasts. .... *P. boldii*

P. atlantica. A soft, filamentous alga forming a brown to dark purple to red mat of short ¾ inch (2 cm) tufts. Branching is lightly dichotomous. Cortication is absent, segments 1–3 diameters long with four pericentral cells. Tetrasporangia are spherical, tetrahedrally divided, and found in swollen, straight branchlets. The plant is commonly found attached to jetty rock or any hard substrate in the lower intertidal zone.

P. boldii. A reddish-brown to almost black species up to ¾ inch (2 cm) long. The alga has well-developed prostrate and erect axes. The branches arise independently of the trichoblasts. The main axes are up to 100–130 µ in diameter, and the segments have a maximum length of 0.5–1 diameter. Six pericentral cells (occasionally seven) are present. The antheridia are distinct in this species, as each antheridial branch has two arms that bear antheridia rather than one, as is common for the species. The alga occurs in bays, usually attached to oyster shells, rock, and wood, and occasionally on the Gulf of Mexico jetties. It is a common plant present throughout the year.

P. denudata. The lower portions of the thallus often brownish with reddish distal branches. The alga is up to 1½ inches (4 cm) long, and prostrate axes are poorly developed or absent. The polysiphonous branches arise in the axils of the trichoblasts. The main axes are up to 350–400 µ in diameter, and the segments have a maximum length of 1–2 diameters. Six pericentral cells (occasionally seven) are present. The antheridial branches occur singly with trichoblasts. This common alga is present in the intertidal zone of jetties throughout the year.

P. echinata. Plants are dark brown, up to 2 inches (6 cm) long, stiff, and robust. The polysiphonous branches arise in the axils of the trichoblasts. The main axes are covered with spinelike branches about 1 mm long. The main axes are up to 400–600 µ in diameter, and the segments have a maximum length of 0.5–1 diameter. Some cortication is present at the bases of the older axes. Four pericentral cells are present. The antheridial branches occur singly with trichoblasts. The algae are present in bays during early spring. They are found drifting, attached to oyster shells, or epiphytic on larger algae.

P. hapalacantha. Plants are densely tufted with filaments 2–6 inches (4–15 cm) long, thick above and thin below. The branching is subdichotomous and lightly corticated. There are four pericentral cells. Trichoblasts are numerous. The plant is found in shallow water attached to hard substrates.

P. havanensis. Plants are reddish-brown, up to 3½ inches (9 cm) long and soft in texture. The polysiphonous branches arise in the axils of the trichoblasts. The main axes are up to 250–300 µ in diameter, and the segments have a maximum length of 2–4 diameters. Four pericentral cells are present. The antheridial branches occur singly with trichoblasts. The alga is collected in late winter and spring from local bays.

P. ramentacea. Plants are brownish-purple tufts 1–4 inches (3–10 cm) long with coarse filaments. There are four pericentral cells, and the primary branches are strongly corticated. Trichoblasts are sometimes present. The tetrasporangia are few in number and found in upper branches. The plant is commonly found in shallow water.

P. subtilissima. An alga that is blackish-purple, up to 2 inches (6 cm) long with very soft texture. The polysiphonous branches arise independently of the trichoblasts. The main axes are up to 100 µ in diameter, and the segments have a maximum length of 2–4 (occasionally up to 8) diameters. Four pericentral cells are present. The species is found throughout the year in bays.

▲ *Polysiphonia boldii* branching.

◄ *Polysiphonia atlantica* branching (pressed specimen).

▲ *Polysiphonia boldii* trichoblasts.

◄ *Polysiphonia boldii* carpospores.

Order Gelidiales
Family GELIDIACEAE
GELIDIUM
⟹ JEL-id-ee-um
G. crinale (Turner) Gaillon
⟹ krin-AH-lee
G. pusillum (Stackhouse) Le Jolis
⟹ puh-SILL-um

Key to the Species
1. Plants large and freely branched .... *G. crinale*
1. Plants very small with a creeping base; the erect and flat branches spatulate, simple, or barely pinnate........................ *G. pusillum*

The plants consist of erect, branched axes that form stiff, wiry tufts. The branches are mostly terete but are typically flattened or compressed in the distal, upward portion. The structure is of uniaxial construction, and each branch ends with a single apical cell and has a distinct broadened trifurcate tip. Internally, the axis consists of a central large-celled medulla and a smaller-celled cortex to the outside. Thick-walled filaments called rhizines are formed as outgrowths of the medullary cells and will grow downward, penetrating between the cells. The presence and position of these rhizines in the medulla may have diagnostic value. The antheridia consist of small, colorless cells and form sori on the surface of the axis behind the apex. The tetrasporangia, which are cruciately divided, are developed in the cortex in the swollen axes at the apices. The two species are separated on plant size and branching.

Typically, *Gelidium* is found in the lower to upper intertidal zone attached to wooden pilings and hard substrates. It may also be observed in bays with low energy attached to shells and human-made structures. It is the most common alga found at the Port Aransas jetties, where it dominates the algal flora near the upper zone of growth. It is found throughout the year but reaches best development in the spring and fall during normal changes of the water temperature.

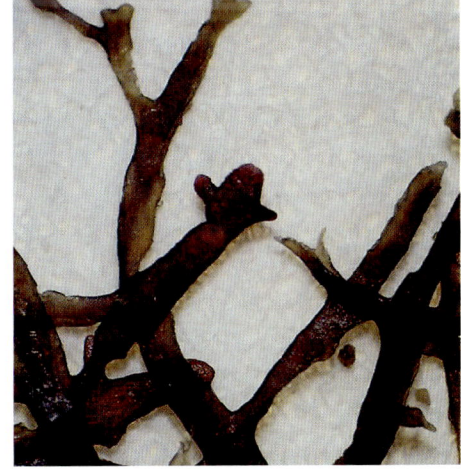

Gelidium crinale branching (close-up).

Gelidium crinale branching (pressed specimen).

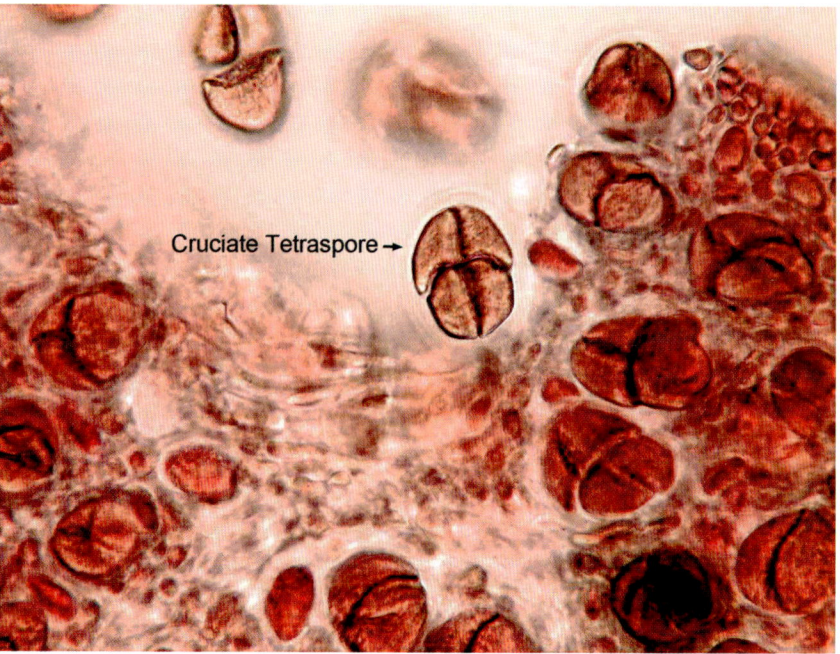

Gelidium crinale cruciate tetraspores.

Order Gigartinales
Family CYSTOCLONIACEAE (HYPNEACEAE)
HYPNEA
▶ HIP-nee-uh

H. cornuta (Kützing) J. Agardh
▶ kor-NOO-tuh

H. musciformis (Wulfen in Jacqu.) J. V. Lamouroux
▶ mus-SEE-for-miss

H. spinella (C. Agardh) Kützing [*H. cervicornis*]
▶ spy-NEL-uh

H. valentiae (Turner) Montagne [*H. cornuta*]
▶ val-en-TEE-ay

The genus contains large, irregularly branched plants with terete axes bearing numerous short, spinelike or stellate branchlets. The structure is uniaxial, and a single large apical cell terminates each branch. A small-celled cortex surrounds the large-celled medulla, which forms a central filament of colorless cells. The sexual plants are dioecious. The antheridia form superficial sori. Cystocarps are urn-shaped to nearly spherical and form prominent swellings on the axes. The zonately divided tetrasporangia are scattered in the outer cortex of slightly swollen ultimate branchlets.

Key to the Species
1. Plants terete with some axes having swollen curved tips. . . .*H. musciformis*
1. Plants lack swollen curved tips. 2
2. Branches 1–4 mm in diameter . . . . *H. valentiae*
2. Branches 0.4–1.5 mm in diameter. 3
3. Branchlets spine- or spurlike; cystocarps produced singly . . . . . .*H. spinella*
3. Branchlets often distinctly star-shaped; cystocarps clustered . . .*H. cornuta*

H. cornuta. This plant is up to 8 inches (20 cm) high, dark green to red with irregular branching. It is commonly bushy and tangled and sometimes found as an epiphyte or a drift alga. The branches have upcurved apices that taper to a point. Branchlets are star-shaped (stellate), which enhances the tangled shape. These may be densely crowded, have up to six points, and are possible dispersal propagules. Tetrasporangia are zonately divided. Cystocarps are urn-shaped to spherical and clustered on side branches.

H. musciformis. The species is extremely variable in color and habit. The color varies from purplish to dark green to reddish-brown. The plant's main axis is up to 8 inches (20 cm) long, is irregularly branched, and has small, spinelike, scattered branchlets. The apices of the principal branches may be strongly curved and thickened to form tendrils. The plant has vigorous growth in areas subject to some wave action; in quiet waters the plant is of a smaller diameter and more fragile. The alga occurs in the deeper algal zones of jetties and bays during the warmer months of the year.

H. spinella. This plant has terete branches that are light brownish to rosy-red that lack stellate branchlets and hook-shaped main branches. The branching is pseudodichotomous below and alternate and cervicorn (like a deer antler) above, loosely covered with spirally arranged simple to branched ultimate branchlets. Hemispherical cystocarps are produced singly. Tetraspores are zonately divided and found in raised sori.

H. valentiae. The plant is up to 10 inches (25 cm) long and varies from purplish to dark green to reddish-brown and is irregularly branched. The apices of the branches are upcurved and taper to a point. The dense branchlets are spine- or spurlike, rarely star-shaped. The alga occurs in the deeper algal zones of jetties and bays during the warmer months of the year. The zonately divided tetrasporangia are found in swollen sori at the base of branchlets. Cystocarps are urn-shaped to spherical and clustered on side branchlets.

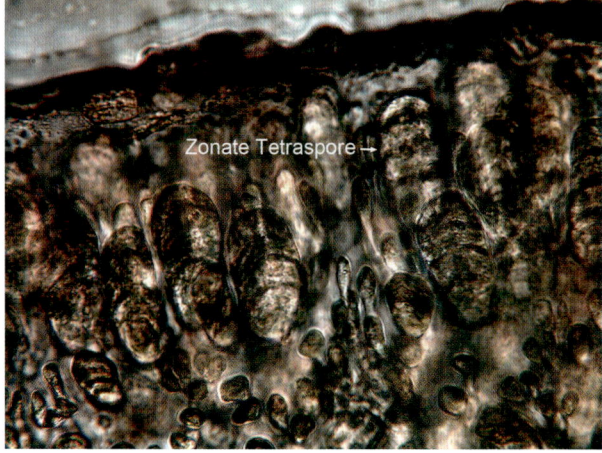

◄ *Hypnea spinella* branching (pressed specimen).

▼ *Hypnea musciformis* zonate tetraspores.

▶ *Hypnea valentiae* habit.

▼ *Hypnea musciformis* hamate tips.

▼ *Hypnea valentiae* carpospores.

▼ *Hypnea cornuta* tangled mass with stellate branchlets.

▼▶ *Hypnea cornuta* stellate branchlets (close-up).

Family SOLIERIACEAE
AGARDHIELLA
Agardhiella subulata (C. Agardh) Kraft et M. J. Wynne [*A. tenera, Solieria tenera*]

⟹ ag-ard-ee-EL-uh sub-yoo-LAH-tuh

Key to the Species
1. Plant with holdfast with many fibrous rhizoids; southern coast . . . . see **Solieria filiformis**
1. Plant thallus axis terete or flattened, radially to subpinnately branched with a simple discoid holdfast; northern coast . . . . . . **A. subulata**

Two species have consistently been known along the Texas coast in the genus *Agardhiella: A. ramosissima* (Harvey) Kylin 1932 and *A. subulata* (C. Agardh) Kraft and Wynne 1979. The taxonomical classification of the genus and the species represented has been highly confusing. Both species (*ramosissima* and *subulata*) were considered *A. subulata*. This is now considered the correct species for the northern Texas shorelines, with *Solieria* f. [*A. subulata*] having a more southern distribution. *Agardhiella subulata* is terete and radially to subpinnately branched but may be flattened at the axis. In addition to differences in their reproductive (cystocarpic) structure, these two taxa are easily separated by their attachment structures. A simple discoid holdfast is found in *A. subulata* with a more complex system with many secondary fibrous rhizoids (rootlike structures) in *S. filiformis.*

A. subulata. This is a large plant, up to 7 inches (18 cm) tall, with a bright reddish-purple color. It has a fleshy texture with terete branches that taper from the center to their bases and apices and may be flattened at the axis. The thallus construction is multiaxial with a medulla consisting of a core of parallel longitudinal filaments. The medulla has an inner core of larger cells and an outer core of smaller pigmented cells. The carposporophytes are embedded in the axes and cause the branches to swell. Tetrasporangia are zonately divided and scattered over the surface of the branches. Plants occur in the lower subtidal zone of jetties. They reach their maximum development during the warmer part of the year.

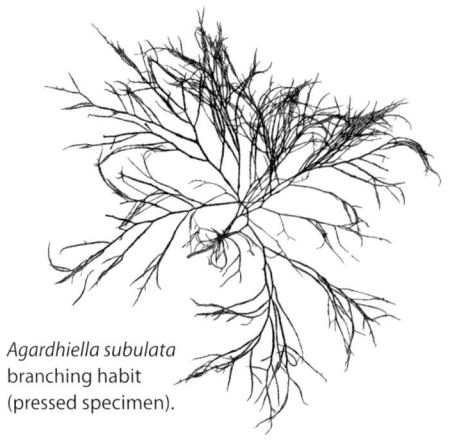

Agardhiella subulata branching habit (pressed specimen).

Agardhiella subulata species showing variation in thallus morphology.

SOLIERIA
Solieria filiformis (Kütz.) P. W. Gabrielson [*Agardhiella tenera*]
⏵ so-li-ER-ee-a fil-ih-FOR-miss

This is a large plant, up to 7 inches (18 cm) tall, with a bright reddish-purple color. It has a fleshy texture with terete branches that taper from the center to their bases and apices. The thallus construction is multiaxial with a medulla consisting of a core of parallel longitudinal filaments. The medulla has an inner core of larger cells and an outer core of smaller pigmented cells. The carposporophytes are embedded in the axes and cause the branches to swell. Tetrasporangia are zonately divided and scattered over the surface of the branches. Plants occur in the lower subtidal zone of jetties. They reach their maximum development during the warmer part of the year.

Solieria filiformis habit.

Solieria filiformis in situ.

Order Gracilariales
Family GRACILARIACEAE
GRACILARIA
Gracilaria tikvahiae McLachlan [*G. foliifera* var. *angustissima*]
⇒ GRASS-il-ar-ee-a ti-KEY-va-hee-ay

Key to the Species
1. Thallus terete throughout with tapering branches . . . . see ***Hydropuntia*** **sp.**
1. Thallus usually compressed at the portion below a forked branch or may be flattened throughout . . . . ***G. tikvahiae***

This species is particularly variable in regard to morphology and growth characteristics. The plant may reach a length of up to 15 inches (37 cm); colors are various, including black, green, and red. The branching is largely in one plane with the axes noticeably compressed, especially in the areas just below where the branches fork. The medulla comprises large, somewhat compressed hyaline (clear) cells and is surrounded by a one- or two-layered cortex of smaller cells. The tetrasporangia are cruciately divided. Spermatangia are found throughout the blade in shallow, concave sori. The plant is present throughout the year in the lower zones of algal growth.

Gracilaria tikvahiae branching habit (pressed specimen).

▼ *Gracilaria tikvahiae* carposporangium.

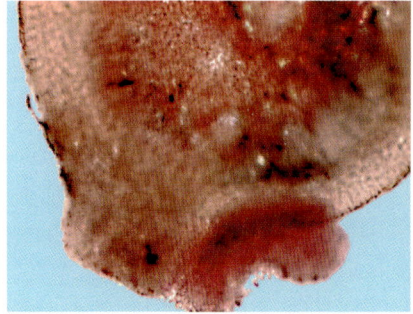

▶ *Gracilaria tikvahiae* in situ.

▼ *Gracilaria tikvahiae* variation in color and branching (close-up).

HYDROPUNTIA
⇒ hy-droh-PUN-tee-uh

H. caudata (J. Agardh) Gurgel et Fredericq [*Gracilaria caudata, G. verrucosa*]
⇒ kaw-DAH-tuh

H. cornea (J. Agardh) M. J. Wynne [*Gracilaria cornea, G. debilis*]
⇒ KOR-nee-uh

Key to the Species
1. Thallus terete throughout with tapering branches . . . . **H. caudata**
1. Thallus usually terete, branched in many planes, lower branching irregularly alternate with cervicorn distal branching . . . . **H. cornea**

H. caudata. The plant has a variable morphology and is up to 24 inches (60 cm) long with a bushy black, brown, green, gray, or pink to red thallus. The axes are continuously terete, branch in many planes, and will taper distally to form fine ultimate axes. The apices are acute. The large hyaline cells of the medulla are continuous with a two- or three-layered cortex of small, rounded-angular cells. The tetrasporangia are cruciately divided. The consistency of the alga is firmly fleshy and adheres well to paper. The plant is found in bays throughout the year but is most common during the warmer part of the year.

H. cornea. This bushy alga varies from green to reddish-brown and is up to 12 inches (30 cm) long. The axes are terete and branched in many planes, lower branching irregularly alternate with cervicorn distal branching. The plant thallus is coarse and does not taper distally or adhere well to paper. A section of the thallus reveals a medulla with large thick-walled cells that gradually grade into smaller surface cells. This common plant is found in bays throughout the year but is best developed during the warmer months.

Hydropuntia caudata habit.

Order Halymeniales
Family HALYMENIACEAE
GRATELOUPIA
⮕ GRA-tee-loop-ee-uh

G. filicina (J. V. Lamouroux) C. Agardh
⮕ fil-ih-SY-nuh

G. pterocladina (M. J. Wynne) S. Kawaguchi et H. W. Wang in Wang et al. [*Prionitis pterocladina*]
⮕ ter-oh-KLAD-ee-nuh

Key to the Species
1. Slimy texture with a lax medulla and scattered reproductive structures . . . .*G. filicina*
1. Rigid texture, a compact medulla, and reproductive structures confined to specific areas of the thallus . . . . *G. pterocladina*

G. filicina. Its color is variable and ranges from a dark red or purple to reddish-brown. The thalli are in clumps or tufted, have a smooth, slimy texture, and are 1–8 inches (2–20 cm) long. The plant has a small holdfast with a short stipe. The flattened (lax) axis runs the length of the plant, with branches growing along the entire length. The branches tend to be more abundant at the top of the thalli. Some populations may exhibit slightly compressed branching; others may have cylindrical, radial branches. And at times both of these characteristics may be seen. The branches may or may not have branchlets. The medulla is composed of periclinal colorless cells and 12 layers of stellate colorless cells. The outer cortex is composed of 4–28 dichotomously divided cell rows that surround the inner cortex of stellate cells. The alga is monoecious. The antheridia form patches of small cells on the thallus surface. Carposporophytes are small and immersed in the axes. The tetrasporangia are embedded in the outer cortex and are cruciately divided.

G. pterocladina. This plant mistakenly appears in previous treatments of marine algae of the Texas coast as at least four different taxa: *Pterocladia capillacea, P. bartlettii, Gigartina elegans,* and, most recently as *Prionitis pterocladina.* The plant is erect, bushy, robust, 2–8 inches (5–20 cm) tall, and is a deep reddish-purple to nearly black. The alga has flattened, pinnately branched axes attached to a fibrous holdfast. The paddle-shaped blades grow from a single apical cell with medullary cells intermingled with thick-walled rhizines, which are most often found concentrated in the center of the blade. The tetraspores are found in the subsurface distal end of the blades and are most often tetrahedrally divided but occasionally may be cruciate or irregular. The alga occurs below the low intertidal zone on area jetties throughout the year.

Grateloupia filicina habit (pressed specimen).

Grateloupia filicina branching and flattened thallus.

HALYMENIA
Halymenia floridana J. Agardh [*bermudensis*]
⏵ hal-ee-MEEN-ee-a flor-ih-DAY-na

This is a gelatinous, foliose (flat-bladed), pinkish to rosy-red plant attached to jetties by discoid holdfasts. The plant gives rise to one or more erect, ovate to suborbicular blades with a wedge-shaped base attached by a short stipe. The apices are rounded. The thallus medulla is composed of densely arranged, mostly longitudinal filaments, some of which bridge the medulla. Tetrasporangia are ellipsoid with cruciately divided spores. The plant is found in deeper water attached to hard substrates. It is rare along the southern coast but common along northern rocky shorelines (especially Galveston Bay).

Halymenia floridana mature thallus in situ. Above is an old thallus; below is a young thallus.

Halymenia floridana young thallus in situ.

Order Rhodymeniales
Family RHODYMENIACEAE
BOTRYOCLADIA
Botryocladia occidentalis (Børgesen) Kylin
⟹ bot-ree-oh-KLAD-ee-uh ok-sih-DEN-tal-iss

The plant has a wiry thallus with branches of spherical-shaped, red-pigmented, grapelike clusters. Branches are arranged alternately to somewhat radially or bilaterally symmetrical. The cortex is three cells thick with a central cavity of clear mucilaginous material. Cystocarps are ovoid to irregular, scattered, and rare. This plant is known from the nearshore waters off the coast of Padre Island National Seashore in about 55 feet (17 m) of water at 7½ Fathom Reef (Texas coast).

Botryocladia occidentalis clusters (close-up; pressed specimen).

RHODYMENIA
Rhodymenia pseudopalmata (J. V. Lamouroux) P. C. Silva
⟹ rho-DY-men-ee-uh soo-doh-pahl-MAY-tuh

The alga is bright reddish to purple with a flattened, dichotomously branched thallus attached by a basal stipe to a small or spreading holdfast. The thallus is up to 4 inches (10 cm) long, the axes are up to ¼ inch (5 mm) wide, and the medulla consists of large, colorless cells surrounded by a layer of small, pigmented cortical cells. It has a multiaxial construction. The plants are dioecious. The antheridia are superficial sori of small, colorless cells. The cystocarps are hemispherical, very prominent, and found spread along the margins and surfaces of the blade. The tetrasporangia are embedded among the cortical cells in the distal portions of the branches and are cruciately divided. It is a common inhabitant of the jetties in the lower portion of algal growth. It may also be found attached to hard substrates in the bays. The alga occurs year-round but reaches maximum development during the middle to late part of the summer. In the bays, it is found only during the hottest part of the year.

Rhodymenia pseudopalmata habit.

Rhodymenia pseudopalmata branching (pressed specimen).

Family CHAMPIACEAE
CHAMPIA
Champia parvula (C. Agardh) Harvey
⇒ CHAMP-ee-uh PAR-voo-luh

The brown to green or rosy-red plant arises from a small discoid holdfast. Plants are erect to prostrate, soft, and branched alternately or suboppositely to irregular. The branches are terete and have a hollow medulla with constrictions at the nodes. Thin, cellular septa transverse the cell at the nodes. The thallus has a central layer of large cells surrounded by a cortical layer of smaller cells. The sexual phase is dioecious. Antheridia are formed in sori found on the thallus surface. Cystocarps are scattered over segments. Tetraspores are tetrahedrally divided. The plant is frequently found in bays and the Laguna Madre during February to July.

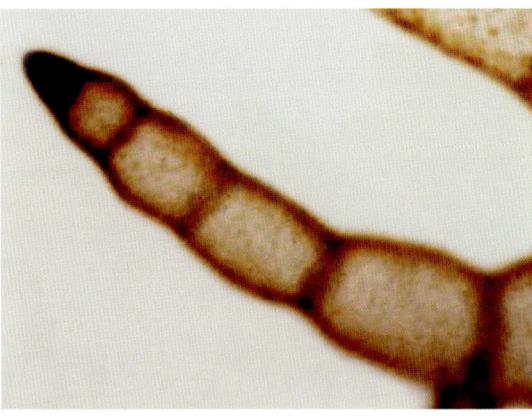

Champia parvula branch morphology (swollen cells).

Family LOMENTARIACEAE
LOMENTARIA
Lomentaria baileyana (Harv.) Farl. [*L. uncinata*]
⇒ loh-men-TAR-ee-a bay-lee-AY-nuh

The plant is small, about ½ inch (1 cm) high, forming a tuft with hollow, cylindrical branches that taper at both ends. The distal portions of the branches are typically curved. The thallus is bright pink to maroon-red, slick with a slimy feel, and may be tangled. Tetrasporangia are tetrahedrally divided and scattered in the subcortical layer. The plant is found attached to rocks or as an epiphyte on seagrass and other algae in calm, shallow water just below the low-tide mark.

LOMENTARIACEAE

Lomentaria baileyana branching habit.

Lomentaria baileyana branches with cystocarps.

Introduction to the Common Brown Seaweeds

Members of this class demonstrate a range of thallus organization extending from simple filamentous representatives to the largest and most complex structures observed in the algae. The simplest forms of brown algae are found in the order Ectocarpales, and they frequently occur as epiphytes on larger plants. The genera are commonly found in intertidal habitats. The species of *Ectocarpus* demonstrate the heterotrichous habit in the majority of cases, with the prostrate branches showing apical growth. Growth of the erect filament is usually by the division of cells within the filament, but in some species the meristematic region is restricted to the base of a terminal hairlike outgrowth. In a few instances, the presence of a terminal meristematic cell has also been described. This group of seaweeds has a characteristic brownish color because the carotenoid pigment (fucoxanthin) masks the remaining pigments (chlorophylls *a* and *c*, other carotenes, and xanthophylls). The pigments are contained in ribbonlike or discoid (lens-shaped) plastids in the cell. Typically, there are pyrenoids (starch-producing structures) associated with the chloroplasts. The reserve food products are mannitol and laminarin. The cell walls contain cellulose, algin, and pectin. Individual cells have a single nucleus. Reproductive cells are motile with two laterally inserted flagella of unequal length present in the majority of local group members. *Dictyota* and *Padina* also have nonmotile reproductive cells.

List of Common Brown Seaweeds

Division Ochrophyta

Class Phaeophyceae

Order Dictyotales
Family DICTYOTACEAE
    Dictyota menstrualis (Hoyt) Schnetter, Hörnig, & Weber-Peukert [*D. dichotoma*]
    Padina gymnospora (Kützing) Sonder [*P. vickersiae*]
    Spatoglossum schroederi (C. Agardh) Kütz.
Order Ectocarpales
Family ACINETOSPORACEAE
    Feldmannia indica (Sond.) Wormersley et A. Bailey [*Giffordia indica, Hincksia duchassaingiana, H. indica*]
    Hincksia mitchelliae (Harv.) P. C. Silva [*Giffordia mitchelliae, Ectocarpus mitchelliae*]

Family CHORDARIACEAE
    Cladosiphon occidentalis Kylin
Family ECTOCARPACEAE
    Ectocarpus rallsiae Vickers [*Giffordia rallsiae, Hincksia rallsiae*]
    E. siliculosus (Dillwyn) Lyngbye [*E. confervoides*]
Family SCYTOSIPHONACEAE
    Petalonia fascia (O. F. Müller) Küntze
Order Fucales
Family SARGASSACEAE
    Sargassum filipendula C. Agardh
    S. fluitans (Børgesen) Børgesen
    S. natans (Linnaeus) Gaillon

Generic Key to the Common Brown Seaweeds

1. Vegetative portions of plants one cell wide throughout; composed of branched filaments without cortication or pseudoparenchymatous organization; reproductive organs terminal. 2
1. Vegetative portions of plants more than one cell wide, at least in part; of various form and pseudoparenchymatous or parenchymatous construction present 4
2. Chloroplasts 1–several; elongate, simple, or forked bands, each with several pyrenoids (1) . . . . **Ectocarpus**
2. Chloroplasts numerous; discoid or rod-shaped with a single pyrenoid; branches arising at acute angles from main axis. 3
3. Plurilocular organs are usually less than 5 diameters long (2) . . . . .**Hincksia**
3. Plurilocular organs are usually more than 5 diameters long. . . **Feldmannia**
4. Plants more or less erect, composed of stems and leaflike thallus usually bearing hollow spherical vesicles (1) . . . .**Sargassum**
4. Plants parenchymatous or filamentous (pseudoparenchymatous); not hollow 5
5. Plants with evidence of filamentous construction with uniseriate branch tips; composed of numerous axial filaments of medulla firmly joined into a pseudoparenchymatous core (4). . . . . **Cladosiphon**
5. Plants parenchymatous, without evidence of filamentous construction (examine in cross section) or of uniseriate branch tips; thalli more or less compressed, linear-strap-shaped to broadly fan-shaped. 6
6. Plants branched; growth at branch tips from a single apical cell; thallus structure throughout of one layer of large medullary cells and a single smaller-celled cortical layer (5) . . . .**Dictyota**

6. Plants forming entire or lobed fan-shaped segments or consisting of a series of 1–several unbranched straplike erect portions 7
7. Plants composed of 1–several unbranched straplike portions (6) . . . . *Petalonia*
7. Plant growth at tips from a marginal row of many apical cells 8
8. Plants more or less fan-shaped, usually calcified; growing margin rolled inward (7) . . . . *Padina*
8. Plants lacking concentric zones, more than ½ inch (1 cm) wide, at times branched (forked) and the growing margin not inrolled . . . . . *Spatoglossum*

Plant Descriptions and Images

Division Ochrophyta

Class Phaeophyceae

Order Dictyotales
Family DICTYOTACEAE
DICTYOTA
Dictyota menstrualis (Hoyt) Schnetter, Hörnig, & Weber-Peukert
[*D. dichotoma*]
⟹ dik-tee-OH-tuh men-strew-AL-lis

This plant consists of flat, brown, dichotomously branched blades with smooth margins, up to 7 inches (18 cm) long, that are attached by a short stipe. The plant grows by means of a single apical cell at the apex of each branch. The blade is three cell layers thick with a single layer of large medullary cells (colorless) with smaller pigmented (with discoid chloroplasts) cortical cells on either side. Hairs are scattered over the surface in prominent tufts. Sexual reproductive structures are found on both surfaces (dorsal and ventral) and appear as small raised characters that will usually have a darker brown to black color. Tetrasporangia are scattered, 1–3 (not in sori), and produce gametes once per month at spring tides. The plant is most often found attached to jetty rocks but may also be established on various hard substrates or as an epiphyte on seagrasses in the coastal bays. Common and most often found as a single attached specimen about 2 feet (60 cm) below the mean low water. The plant may be collected throughout the year but is most abundant during the warmer parts of the year.

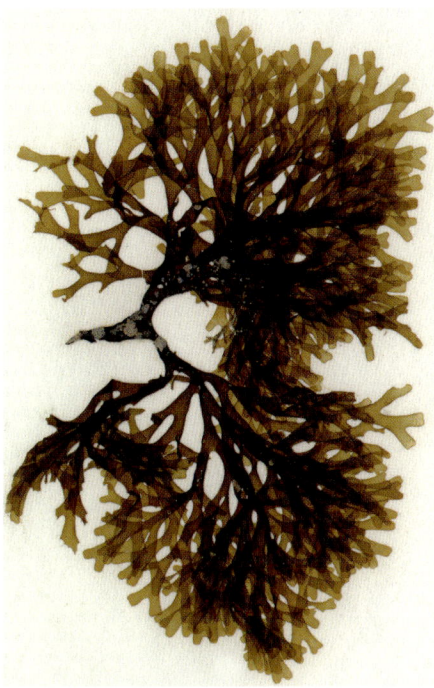

▲ *Dictyota menstrualis* branching morphology (pressed specimen).

▲ *Dictyota menstrualis* single apical cell at branch apex (recently divided).

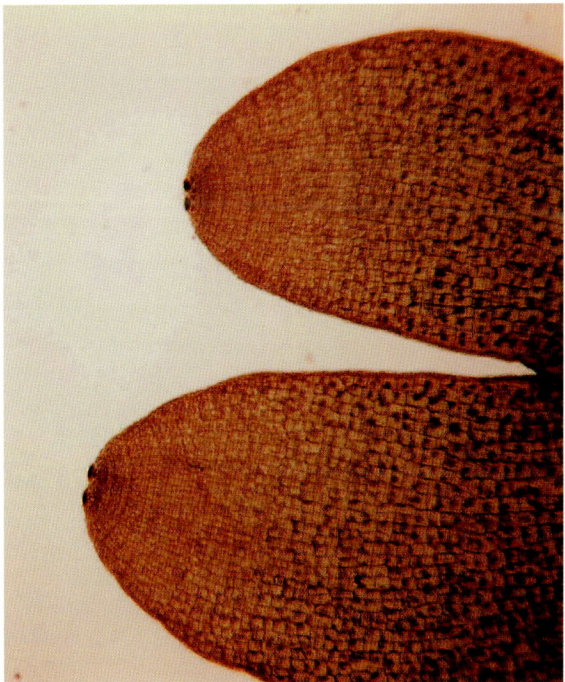

▶ *Dictyota menstrualis* dichotomous branching.

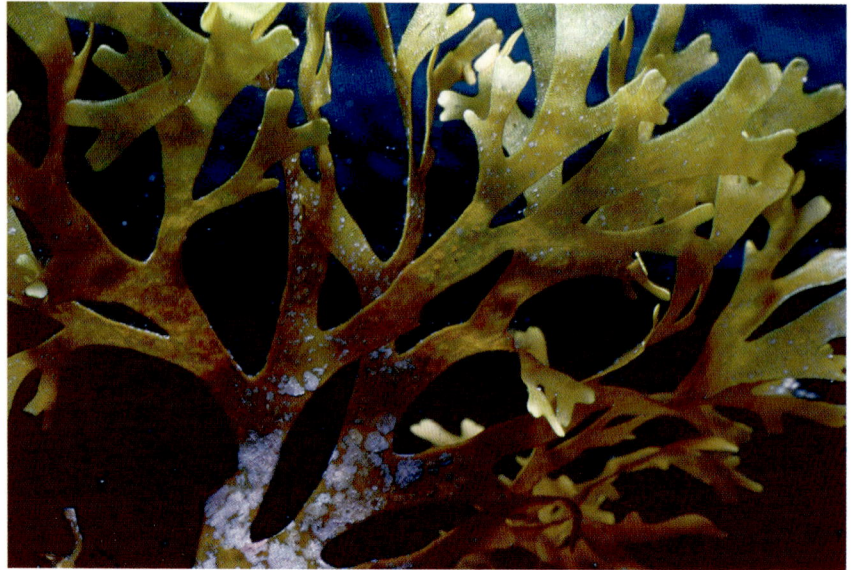

Dictyota menstrualis in situ.

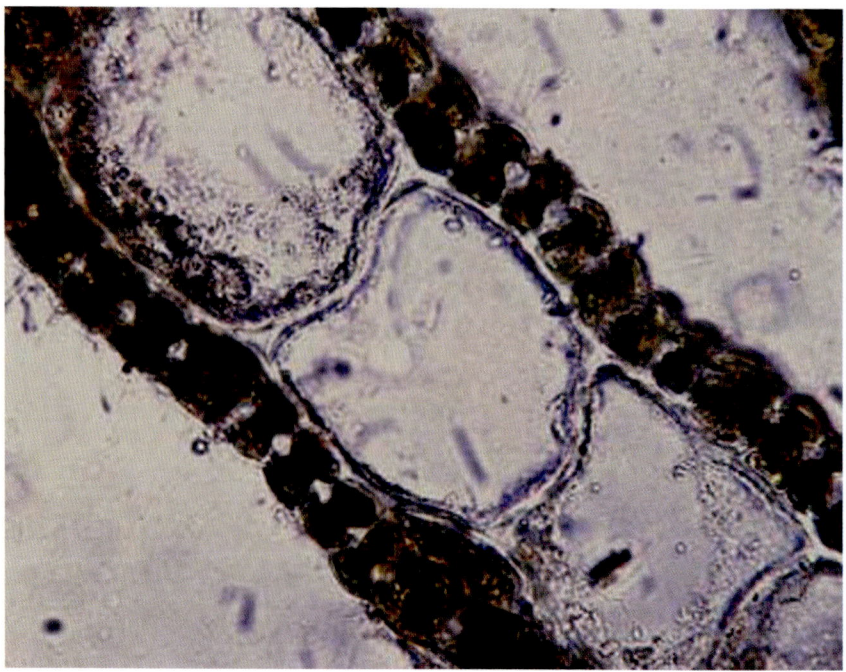

Dictyota menstrualis tristromatic (three layer) thallus (microscopic lateral section).

PADINA
Padina gymnospora (Kützing) Sonder [*P. vickersiae*]
> pah-DY-nuh jim-noh-SPOR-a

This alga is brown in color but may become whitish with the deposit of limestone over the blade's surface. The flat, fan-shaped blade, up to 8 inches (20 cm) long, is marked by zones of concentric rows of hairs located parallel to the margins. Growth is marginal along the distal edges of the blades, which are circinately inrolled (rolled inward) to protect the meristematic cells. The fan may become split as the plant ages and, therefore, may take on a more ribbon-like appearance. The plant is attached at the lower depths of algal growth to a hard substrate, usually jetty rock, by a disk-shaped holdfast. In the bay, the plants are sometimes found attached to rocks, shells, or other hard material and, in these areas of lower energy, tend to be less damaged than the jetty plants. Reproductive structures (sori) appear in one or two bands as darkened bodies between the concentric rows of hairs, usually on the ventral surface. The reproductive structures (sori) may be either naked antheridia (male) or oogonia (female) covered with sterile cells or cruciately divided tetrasporangia covered with sterile cells. The plant is found throughout the year but is common only during the warmest periods.

Padina gymnospora habit.

▲ *Padina gymnospora* antheridia.

Padina gymnospora circinately inrolled margin.

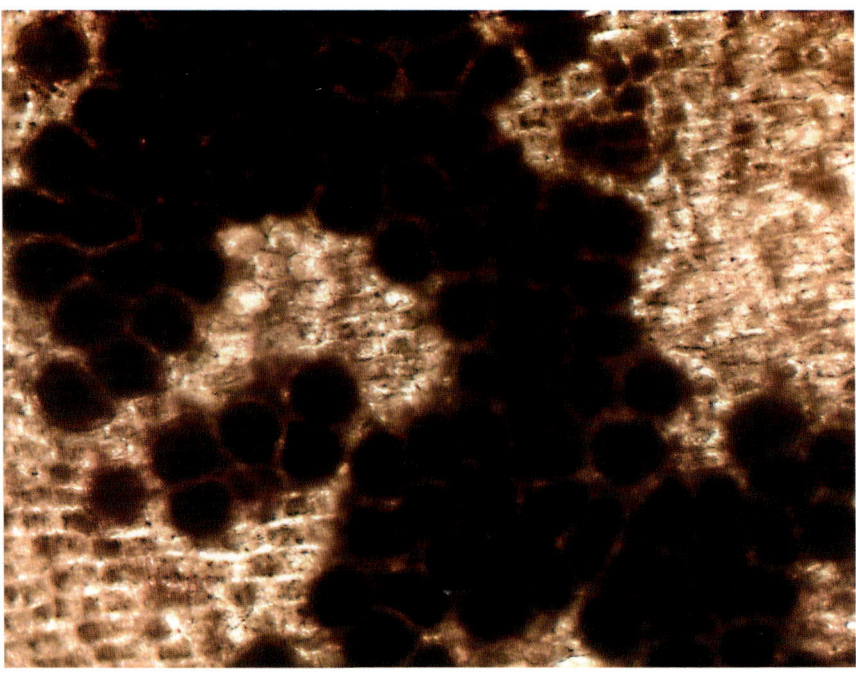

Padina gymnospora reproductive sporangia.

SPATOGLOSSUM
Spatoglossum schroederi (C. Agardh) Kütz.
⏵ spat-oh-GLOSS-um SHROW-der-eye

The plant is somewhat stalked with a flat, strap-shaped, and irregularly branched thallus. It may reach a height of 8 inches (20 cm) or more. The apices are rounded with typically pointed marginal teeth at irregular intervals. It is bright golden-brown when young and ages to a darker brown with a thicker thallus that may become iridescent. Upon collection, it may turn a blue-green color and stain other algae with which it comes in contact. Tetrasporangia are found scattered on both surfaces of the blade. The plant is found on jetties in the lower regions of algal growth or attached to hard substrates in southern bays and estuaries. This alga occurs only during the warmer (summer) parts of the year.

Spatoglossum schroederi habit.

Spatoglossum schroederi habit.

Order Ectocarpales
Family ACINETOSPORACEAE
FELDMANNIA
Feldmannia indica (Sond.) Wormersley et A. Bailey [*Giffordia indica, Hincksia duchassaingiana, H. indica*]
⟹ FELD-man-ee-uh IN-dih-kuh

This plant is a delicately branched, filamentous alga that forms brown tufts up to 8 inches (20 cm) long. The plants are often misidentified as *Ectocarpus*, but the reproductive structures are different in the two genera. *Feldmannia* may be readily recognized by their greatly elongated plurilocular sporangia, which are usually more than 5 diameters long and have blunt apices. The small, epiphytic plants are found on the jetties or attached to any hard material such as rocks, shells, boats, pilings, and bottles in bays and estuaries. It is a pioneering species and is often the first to establish on a disturbed area. The alga occurs all year but reaches its maximum coverage in late winter and spring.

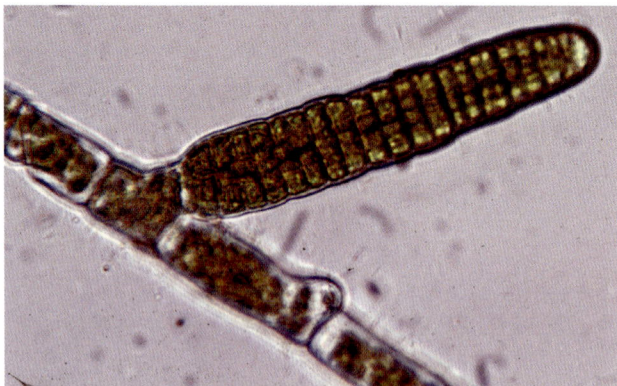

Feldmannia indica branching and chloroplast morphology.

Feldmannia indica epiphyte on red alga.

HINCKSIA
Hincksia mitchelliae (Harv.) P. C. Silva [*Giffordia mitchelliae, Ectocarpus mitchelliae*]
⇒ HINK-see-uh my-CHEL-ee-ay

The plant is a delicately branched filamentous alga that forms yellow-brown tufts or masses up to 8 inches (20 cm) long. The longer filaments lack colorless hairs, are much branched, and sometimes are tangled together. The chloroplasts are discoid in shape with pyrenoids. The plurilocular sporangia are short, usually less than 5 diameters long, and have blunt apices. The unilocular organs are oblong in profile. This is a common species that occurs throughout the year on rocks, oyster shells, and wooden pilings and as an epiphyte on larger algae and seagrasses.

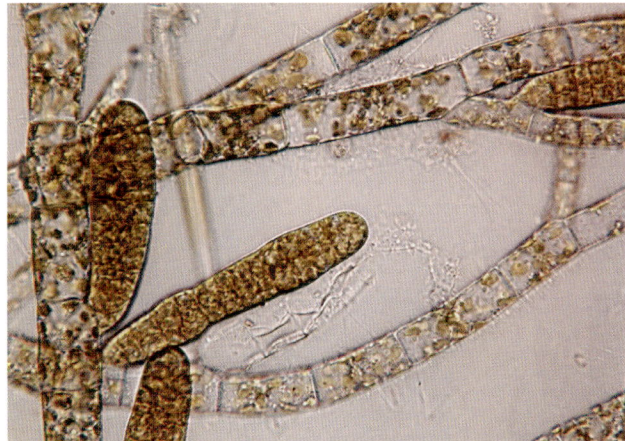

Hincksia mitchelliae plurilocular organ (rounded).

Hincksia mitchelliae in situ.

Family CHORDARIACEAE
CLADOSIPHON
Cladosiphon occidentalis Kylin
kla-do-SY-fon ok-sih-den-TAY-liss

The plant consists of cylindrical, many-branched filaments up to 1/16 inch (2 mm) wide and up to 8 inches (20 cm) long. It is golden-brown to dark brown in color. The construction is pseudoparenchymatous in that the thalli are composed of many filaments that result in a somewhat flat, bladelike morphology. These filaments are covered with fine, colorless hairs and are enclosed in mucilage, which gives the plant a slippery texture. Growth is primarily through intercalary cell divisions that are restricted to zones at the bases of the colorless hairs (trichothallic growth) in the cortex. Unilocular sporangia are produced at the bases of the peripheral photosynthetic filaments. The plurilocular organs are formed at the tips of the outer filaments and are branched, pluriseriate structures. They are epiphytic upon seagrasses, most notably *Thalassia,* and also on coarse seaweeds found in bays and estuaries. The plant appears in early spring, grows and develops very rapidly, but is somewhat rare.

Cladosiphon occidentalis epiphyte on *Thalassia.*

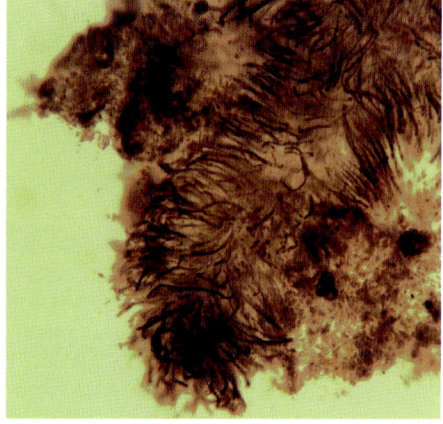

Cladosiphon occidentalis pseudoparenchymatous structure (close-up).

Family ECTOCARPACEAE
ECTOCARPUS
  ⇛ ek-toh-KAR-pus
E. rallsiae Vickers [*Giffordia rallsiae, Hincksia rallsiae*]
  ⇛ RAL-see-uh
E. siliculosus (Dillwyn) Lyngbye [*E. confervoides*]
  ⇛ sil-ee-koo-LOH-sus

Key to the Species
1. Chloroplasts discoid and numerous . . . . ***E. rallsiae***
1. Chloroplasts parietal ribbons embedded with pyrenoids . . . . .***E. siliculosus***

E. rallsiae. This plant is a minute, microscopic, and tufted alga that rarely reaches 1 inch (2 cm) in length. The short plurilocular organs (sporangia) with pointed apices are very distinctive. *Ectocarpus*, like many other brown algae, has a complex life cycle in which alternation of generations occurs. Plurilocular organs (sporangia) are the only type of reproductive body observed in the Texas coastal area and are variable in form but are usually greatly elongated, sessile (or on short stalks), and usually terminate in a colorless hair. This alga is commonly found in winter and early spring.

E. siliculosus. This plant appears as light yellow to delicate brown tufts of seaweed attached to jetty rock or epiphytic on larger algae and seagrasses. The plant is composed of branched, uniseriate filaments with cells containing parietal chloroplasts that are most often ribbons (band-shaped) embedded with pyrenoids. The thallus is small, only about 6–7 inches (16 cm) long. This alga is common in winter and early spring.

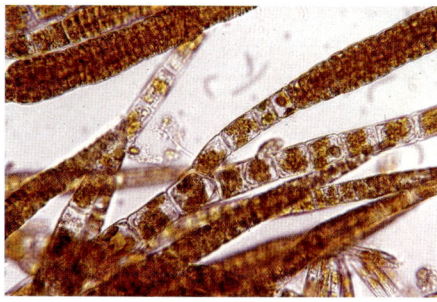

Ectocarpus siliculosus plurilocular sporangia.

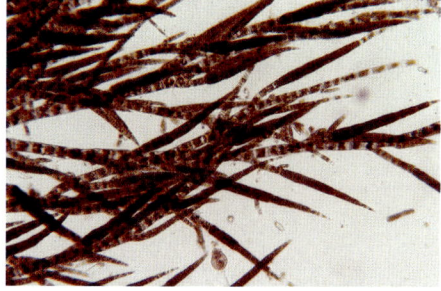

Ectocarpus rallsiae plurilocular sporangia.

Family SCYTOSIPHONACEAE
PETALONIA
Petalonia fascia (O. F. Müller) Küntze
⏩ pet-uh-LOH-nee-uh fash-ee-AH

This small alga has flat, brown, lanceolate-shaped blades, 1–1½ inches (2–2.5 cm) in width, with a height of nearly 8 inches (20 cm). Blade growth is by diffuse cell divisions. The surface of the alga bears colorless hairs. The blade tapers to a short, terete stipe that attaches to jetty rocks by a basal holdfast where branching into multiple blades occurs. The plant is sometimes roughly torn by wave action, and the tip, which is normally acute in shape, may be lacking entirely. Plurilocular organs, the only type of reproductive structures observed in this alga along the Texas coast, are densely packed into sori that cover large portions of the distal aspect of the blade. The result is substantial areas of the blade appearing lighter in color or translucent as the sporangia are discharged into the water. The plant occurs only during the colder months, sometimes only a few weeks of the year. *Petalonia* may also be found attached to hard substrates in bays.

Petalonia fascia habit (pressed specimen).

Petalonia fascia in situ.

Order Fucales
Family SARGASSACEAE
SARGASSUM
⇨ sar-GAS-um
S. filipendula C. Agardh
⇨ fil-ih-PEN-dyoo-luh
S. fluitans (Børgesen) Børgesen
⇨ FLOO-ih-tanz
S. natans (Linnaeus) Gaillon
⇨ NAY-tanz

The plants resemble higher plant construction in organization. They are light to dark brown with a rootlike holdfast (sometimes missing) with branches that resemble leaves of higher plants. These leaves are slender with a midrib and serrated margins and are attached by a petiole. The pelagic species has not been documented to reproduce sexually. These floating forms (pelagic) are found cast up on the shorelines of Gulf and bay beaches. *Sargassum fluitans* vesicles (hollow air bladders for floatation) are spineless, whereas *S. natans* vesicles have a sharp spine on the distal tip of the air bladder. During spring, these two plants are found washed ashore in great quantities.

Sargassum filipendula has vesicles that may have an apical spine or hook. Reproductive structures are found at the point of branching and resemble a cluster of spines. It is attached to Gulf of Mexico jetty rock in areas of high-energy wave action.

Key to the Species
1. Plants attached by a holdfast to jetty rock; vesicles sometimes with a terminal hook or spine. . . . . **S. filipendula**
1. Plants pelagic and lacking holdfast 2
2. Leaves 1/16 inch (1–2 mm) wide with very fine teeth; vesicles with a spinelike projection . . . . **S. natans**
2. Leaves 1/8 inch (3–4 mm) wide with broad teeth; vesicles without a spinelike projection. . . . . **S. fluitans**

S. filipendula. The brown thallus is upright, tough, and leathery with 1–several main axes that are smooth and lack spines. The blades are long with a distinct midrib and margins that are irregularly toothed. The air bladders are spherical to ¼ inch (5 mm) in diameter, numerous, and often with a single spine at the apex. The plant is strongly attached to Gulf jetty rock and other hard substrates by a conical to lobed holdfast. The plant is found just below the low-tide level during all seasons.

S. fluitans. This pelagic species is easily recognized by numerous, short-stalked, lanceolate leaves ¼ inch (5 mm) wide and ¾–1½ inches (2–6 cm) long with pointed teeth that have a wide base. The main axes may be smooth or have spines. The spherical air bladders are ¼ inch (5 mm) in diameter, numerous, and located one or two at the base of each blade. These vesicles lack a spinelike projection at the apex. The holdfast is lacking, as the thallus is unattached. These plants are found floating in large masses or rafts and will wash ashore on Gulf beaches in great quantities, most often during the spring and summer.

S. natans. This brown alga has narrow lanceolate leaves that are only ⅛ inch (3–4 mm) wide and have marginal teeth that are coarsely pointed and slender. The spherical air bladders are numerous, about ¼ inch (5 mm) in diameter, and have a prominent spinelike apical projection. The holdfast is lacking, as the species is pelagic. These plants are found floating in large masses or rafts and will wash ashore on Gulf beaches in great quantities most often during the spring and summer.

Sargassum fluitans in situ.

Sargassum natans pressed.

Sargassum filipendula pressed.

Introduction to the Common Green Seaweeds

The marine green seaweeds contain some of the most familiar plants of the intertidal community. Some of the species may extend their range into estuaries and salt marshes. Their bright green color makes them an obvious part of the shoreline flora, especially when present in quantity. At times the algae will form prominent zones. Many representatives show their best growth in pools at all levels along the shore, whereas others are situated in habitats that are exposed to air during low tide, and many prefer sandy conditions.

The grass-green color of this group is due to large amounts of the green pigment chlorophyll, which is the pigment found in all plants and necessary for the process of photosynthesis. In this process, the plants produce a simple sugar (food) and oxygen as a by-product, which we breathe. Members of the division Chlorophyta range in size from microscopic to individual plants that may reach lengths of 20 or more inches (50 cm). In the warm-water species found along the Texas coast, the diversity in form and size makes this group of algae one of the most attractive and interesting of all marine seaweeds.

In the majority of multicellular green seaweeds, the plant form is based on a simple filament that may either become branched or develop into a flat sheet. Important exceptions to this are seen in the siphonaceous (hollow tube) construction of the genera *Derbesia, Bryopsis,* and *Codium* and the coenocytic genus *Caulerpa.* In addition, some species are encrusted with calcium carbonate material that may or may not be segmented with flexible joints (geniculate). Many of the plants cannot be identified to species without the use of microscopic characters; however, the casual beach visitor can identify many members of the genera simply on the basis of shape, size, and location.

List of Common Green Seaweeds

Division Chlorophyta

Class Ulvophyceae

Order Ulvales
Family ULVACEAE
    Ulva clathrata (Roth) C. Agardh [*Enteromorpha clathrata, E. muscoides, E. ramulosa*]
    U. fasciata Delile
    U. flexuosa Wulfen [*Enteromorpha flexuosa, E. lingulata*]
    U. lactuca Linnaeus
    U. prolifera O. F. Müller [*Enteromorpha salina, E. prolifera*]

Order Cladophorales
Family CLADOPHORACEAE
    Chaetomorpha brachygona Harvey
    C. gracilis Kützing
    C. linum (O. F. Müller) Kützing
    Cladophora albida (Ness) Kützing [*C. glaucescens*]
    C. dalmatica Kützing [*C. luteola*]
    C. ruchingeri (C. Agardh) Kützing
    C. vagabunda (Linnaeus) van den Hoek [*C. fascicularis*]

Order Bryopsidales
Family BRYOPSIDACEAE
    Bryopsis pennata J. V. Lamour.
    B. plumosa (Hudson) C. Agardh
Family CODIACEAE
    Codium taylorii P. C. Silva
Family CAULERPACEAE
    Caulerpa mexicana Sond. ex Kütz. [*C. crassifolia, C. c.* f. *mexicana*]
    C. prolifera (Forssk.) J. V. Lamour.

Family UDOTEACEAE
    Halimeda incrassata (J. Ellis) J. V. Lamouroux
    Penicillus capitatus Lamarck
Order Dasycladales
Family DASYCLADACEAE
    Batophora oerstedii J. Agardh
Family POLYPHYSACEAE
    Acetabularia crenulata J. V. Lamouroux
    A. farlowii Solms-Laubach

Generic Key to the Common Green Seaweeds

1. Plants encrusted with calcium carbonate............................2
1. Plants not encrusted with calcium carbonate 4
2. Plants distinctly segmented, with flexible uncalcified joints (1)... ***Halimeda***
2. Plants not segmented and lacking flexible uncalcified joints 3
3. Plants lack a disk at apex and terminated by a tuft of dichotomously branched filaments (2) .... ***Penicillus***
3. Plant with a simple stipe terminated by disk-shaped caps toward the apex, calcified and bright green in color......... ***Acetabularia***
4. Plants filamentous or membranous (1) 5
4. Plants neither filamentous nor membranous; stoloniferous or coenocytic...7
5. Plants membranous or tubular (4)...................................6
5. Plants filamentous or of more complex branching.....................8
6. Plant membranous or tubular; thallus two cells thick (5) .... ***Ulva***
6. Plants coenocytic and/or stoloniferous..............................7
7. Plants without a distinct differentiation into stolons and erect portions; inflated utricles (6) .... ***Codium***
7. Plants stoloniferous with an erect thallus of various forms; internal portions with trabeculae............................ ***Caulerpa***
8. Plants with stiff, unbranched filaments attached by a basal holdfast (5).. ***Chaetomorpha***
8. Plants filamentous and complex branching...........................9
9. Plants with whorled branches (8)......................... ***Batophora***
9. Plants with branching other than whorled..........................10
10. Plants with ultimate branches with cross walls at the base; distal branching (9).. ***Cladophora***
10. Plants with fronds triangular to lanceolate, light to somewhat olive-green, with branchlets generally in two rows; plants erect, tufted, and featherlike .... ***Bryopsis***

Plant Descriptions and Images

Division Chlorophyta

Class Ulvophyceae

Order Ulvales
Family ULVACEAE
***ULVA* Sea Lettuce**
- UL-vuh

U. clathrata (Roth) C. Agardh [*Enteromorpha clathrata, E. muscoides, E. ramulosa*]
- klath-RAH-tuh

U. fasciata Delile
- fash-ee-AY-tuh

U. flexuosa Wulfen [*Enteromorpha flexuosa, E. lingulata*]
- fleks-yoo-OH-suh

U. lactuca Linnaeus
- lak-TOO-kuh

U. prolifera O. F. Müller [*Enteromorpha salina, E. prolifera*]
- pro-LEEF-er-uh

These plants are bright green and may be identified by three different forms: several long, narrow strips; a single-lobed blade; or a hollow tube. The membranous blades are either two cell layers thick (distromatic) or a hollow tube that is one cell thick. Branching occurs near the basal holdfast. Some of the cells of the blade may serve as reproductive structures. When the reproductive cells are released from the plant, the area remains empty and will appear white. In addition, plants that have been overexposed to light may lose their chlorophyll and turn white. The species are plastic and, therefore, extremely variable in differing environmental surroundings. Plants are found attached to the jetties in the upper zone of algal growth. Occasionally they may be found attached to rocks and pilings in bays. *Ulva* is found throughout the year but prefers the medium temperatures associated with transitional seasonal periods such as those found in the spring and fall.

Key to the Species
1. Thallus sheetlike 2
1. Thallus tubular, hollow, or (if flattened) with hollow margin 3
2. Plants divided into narrow segments less than ½ inch (1.5 cm) wide . . . . ***U. fasciata***

2. Plants simple or with broad, irregular lobes . . . . ***U. lactuca***
3. Plant with few branches. . . . . ***U. flexuosa***
3. Plant with numerous branches. 4
4. Branching abundant; branches successively smaller than the main axis . . . .
. . . . ***U. prolifera***
4. Branching diffuse along the main axis . . . . ***U. clathrata***

U. clathrata. The bright green thallus is up to 1 foot (30 cm) long with blades up to ⅛ inch (2.5 mm) wide and usually tapering toward the base. The alga has abundant branches that are tubular and hollow with walls one cell thick (monostromatic). Each order of branching is smaller than the preceding. Branchlets are short and spikelike. The cells are arranged in distinct longitudinal series throughout the plant. The rootlike rhizoids are delicate and form a tight weave of attachment. This common species is found attached to shallow hard substrate or other marine plants as an epiphyte in the intertidal zone on rocky jetties and in bays. The alga occurs throughout the year, but maximum development takes place in winter and early spring.

U. fasciata. The bright green thallus is thin, sheetlike, up to 1 foot (30 cm) long, and is divided into linear segments. Each blade is two cells thick (distromatic), less than ¾ inch (1.5 cm) wide, and with smooth or slightly ruffled edges. The base of the blade is wedge-shaped with short rhizoids. The alga is found in the intertidal zone attached to Gulf jetties but may be found in deeper water. It also is associated with areas with high nutrient levels. This common species is present throughout the year but has maximum growth in early summer.

U. flexuosa. The light green plant, which grows in clusters or tufts, is up to 8 inches (20 cm) long and has blades up to ¾ inch (1.5 cm) wide. The alga is generally unbranched but is occasionally divided at the base into two or three similar axes. The tubular, hollow blades taper to a filiform stipe below and are inflated distally. The blade walls are one cell layer thick and are arranged in regular longitudinal rows below but are irregularly set distally. Lateral branchlets are short and narrow. The rhizoids are delicate and form a tight pad of attachment. The alga is observed in shallow pools, attached to high intertidal jetty rock, and in coastal bays. This common species is present from winter to early summer.

U. lactuca. The bright green, thin thallus is up to 8 inches (20 cm) long. It is sheetlike and varies from a simple blade to being broadly lobed. The blade is two cells thick, very delicate, and is often perforated. It may also be ruffled, especially along the margins. The rhizoids are short. The alga is observed in shallow pools, attached to intertidal jetty rock, and in coastal bays. It is often found in water with high nutrient levels. This common species is present from winter to early summer.

U. prolifera. This yellow-green plant is up to 1 foot (30 cm) long and has blades up to ⅜ inch (5 mm) wide. The alga has abundant branches that develop along the entire length of the main axis. The blades are hollow with walls a single cell thick. The cells are arranged in longitudinal rows in the younger parts but become less orderly with maturity. The holdfast is a pad of tightly woven rhizoids. The alga is observed in shallow pools, attached to intertidal jetty rock, and in coastal bays during winter and early spring.

▲ *Ulva fasciata* habit (pressed specimen).

◄ *Ulva lactuca* (pressed specimen).

◄ *Ulva flexuosa* showing tubular structure with few or no branches.

ULVACEAE

▶ *Ulva lactuca* habit.

◀ *Ulva fasciata* habit.

▼◀ *Ulva lactuca* distromatic (two layers) thallus.

▼ *Ulva lactuca* surface cells.

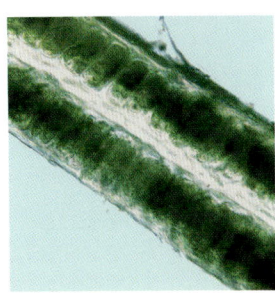
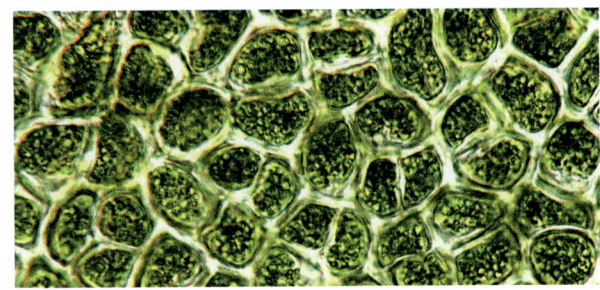

Ulva prolifera abundant branching along hollow thallus.

Order Cladophorales
Family CLADOPHORACEAE
CHAETOMORPHA
⟹ key-toh-MOR-fa
C. brachygona Harvey
⟹ bra-kee-GOH-na
C. gracilis Kützing
⟹ GRASS-il-is
C. linum (O. F. Müller) Kützing
⟹ LY-num

Chaetomorpha means "hair-shaped." The cylindrical filaments are somewhat stiff and curled or twisted, having the texture of the coarse hair of a horse's tail. The plants are traditionally assigned to taxa on the basis of cell diameters, cell length–to–width comparisons, and the presence or absence (and morphology) of basal holdfast cells. The vegetation consists of unbranched, uniseriate (single row of cells) filaments (100–150 µ in diameter) with cells of maximum length of one or two times the cell's diameter. The species is attached by single basal holdfast cells. The multinucleate cells contain a reticulate (netlike) chloroplast that may become fragmented into several smaller pieces. Numerous pyrenoids are present in each cell. Growth proceeds by diffuse cell divisions. The plant occurs on the jetties in the upper zone of algal growth with the filaments attached to rocks either singly or more often in clusters. The jetty type occurs as straight, unbranched filaments reaching a length of up to 8 inches (20 cm). In the bay, the unbranched filaments are found entangled among other algae and seemingly have no holdfast. These filaments may reach a length of 12 inches (30 cm) and may exhibit an intertwined morphology. Both forms are more common during the late spring and summer but may also be found growing during the remainder of the year.

Key to the Species

1. Plants generally epiphytic, filaments under 100 μm in diameter... *C. gracilis*
1. Plants unattached or on hard surfaces, filaments usually over 100 μm in diameter...2
2. Filaments usually softer, tangled in other species; cells about as long as broad with thickened cell walls (1)................ *C. brachygona*
2. Filaments rather stiff; cell walls not unusually thick; cell length greater than one-half cell diameter after division................ *C. linum*

C. brachygona. This green plant is a mass of entangled filaments that are fine and soft. The unbranched filaments are 6 inches to 5 feet (15–150 cm) long with cylindrical to slightly swollen cells that may be straight or curled. This alga is found in both brackish water and saltwater. It is entangled in seagrass beds or in intertidal waters of bays and estuaries. The plant grows best during the warmer months.

C. gracilis. The thallus is yellow-green filaments that are fine and stringlike up to 6 feet (2 m) long. The unbranched filaments are loosely attached by basal cells that easily break. They may be straight, curled, or twisted. The plants grow in the intertidal zone as a filamentous mass attached to hard surfaces. They are often associated with nutrient-rich water. The plant grows best during the warmer months.

C. linum. This yellow-green plant consists of unbranched filaments with cells of maximum length of 1–2 diameters. The multinucleate cells contain a reticulate chloroplast. The plant may form a mass up to 3 feet (1 m) high and 6 feet (2 m) wide. This growth is usually associated with areas of high nutrients. This common alga grows during the warmer months attached to jetties and in bays in the intertidal zone.

CLASS ULVOPHYCEAE

Chaetomorpha linum in situ (close-up).

Chaetomorpha linum attached to jetty rock.

Chaetomorpha gracilis in situ.

Chaetomorpha gracilis barrel-shaped cells.

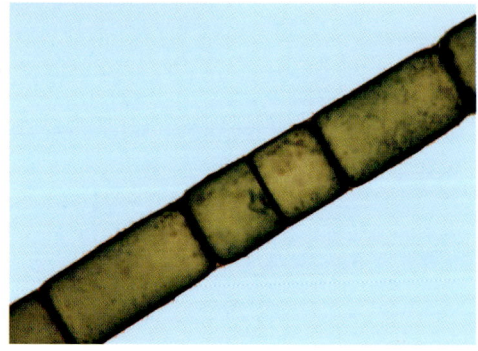

CLADOPHORA
➭ kla-do-FOR-a
C. albida (Ness) Kützing [*C. glaucescens*]
➭ AL-bi-da
C. dalmatica Kützing [*C. luteola*]
➭ dal-MAT-ih-kuh
C. ruchingeri (C. Agardh) Kützing
➭ roo-CHING-er-ee
C. vagabunda (Linnaeus) van den Hoek [*C. fascicularis*]
➭ vag-uh-BUN-duh

These pale to grass-green, branched, filamentous algae are easy to identify to genus because of the typical branching that occurs at the distal (from the holdfast) end of the cell. Also, it is the only green, actively branched filamentous form among the green algae along the Texas coast. However, it is difficult to assign specimens to a particular species because of the variability within the group. The multinucleate cells have a reticulate (netlike) chloroplast containing several pyrenoids. The ratio between apical and intercalary growth is an important taxonomic measure, although it is influenced by water movement in many species. Apical growth is often expressed in agitated waters, and intercalary growth, in quiet waters. This alga grows attached to the granite rocks in the upper zone of algal growth on both the channel and surf sides of jetties. It grows well in rough waters but will not be as bushy as when it grows in calm waters. It is found from late fall through early spring and reaches its maximum development only when the water is relatively cool.

Key to the Species
1. Plant sparingly branched; maximum of one branch per node . . . . *C. ruchingeri*
1. Plant abundantly branched 2
2. Ratio of apical cell diameter to that of the main branch diameter 4 or greater . . . . ***C. vagabunda***
2. Ratio of apical cell diameter to that of the main branch diameter equal to or less than 4 3
3. Plant with acropetally organized, terminal branched system with apices tapering . . . . **C. albida**
3. Plant with predominantly intercalary growth with maximum diameter of apical cells to 32 μm . . . . ***C. dalmatica***

C. albida. The plants are up to 6 inches (15 cm) long and dark green. The thallus is distinctly acropetally organized with the branchlets unilaterally inserted on the main axis. The main filaments are 30–60 μm in diameter with cylindrical cells 2–7 diameters long. Vegetative reproductive cysts (akinetes) are common. The filaments are attached to hard substrates by fine, short rhizoids in the upper intertidal zone during winter and early spring.

C. dalmatica. The bright to dark green plants are up to 8 inches (20 cm) long and are acropetally organized with the branchlets occasionally unilaterally inserted on the main axes. The main axes are 85–150 μm in diameter. This common plant is present throughout the year in bays and estuaries.

C. ruchingeri. The dark green plants are up to 2 inches (5 cm) long and grow largely by intercalary growth and are very sparingly branched. The main axes are up to 200 μm in diameter. The species is found throughout the year attached to Gulf jetties in the intertidal zone.

C. vagabunda. The plants are up to 1 foot (30 cm) long and green to grass-green. The thallus is distinctly acropetally organized with branchlets unilaterally inserted on the main axis. The main axes are 200–300 μm in diameter. The branching is pseudodichotomous below and unilateral above. The cellular division is intercalary, not apical. Branchlets may be slightly constricted at the junction of the main filament. This common species is found throughout the year in the intertidal zone on Gulf jetties and bays.

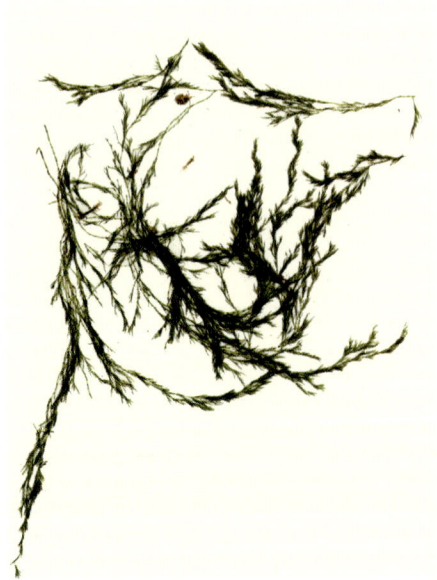

Cladophora vagabunda branching (pressed specimen).

CLADOPHORACEAE

Cladophora albida branching habit.

Cladophora vagabunda branching (pressed specimen).

Order Bryopsidales
Family BRYOPSIDACEAE
BRYOPSIS
⇒ bry-OP-sis

B. pennata J. V. Lamour.
⇒ pen-AY-tuh

B. plumosa (Hudson) C. Agardh
⇒ plum-OH-suh

These plants consist of pinnately branched, hollow (coenocytic) tubes attached by rhizoids (rootlike structures). Plants are dark green and grow as soft tufts of erect branches. The branches somewhat resemble miniature pine trees and reach a height of up to 2¾ inches (7 cm). The thallus contains many nuclei and has discoid-shaped chloroplasts that contain pyrenoids. The lateral branchlets may become segregated from the main axis by a basal membrane and eventually function as gametangia. The plant occurs commonly in protected areas of the uppermost zone of algal growth on jetties and occasionally attached to rocks and other debris in shallow areas of bays. It is a warm-season plant that appears in late spring and persists throughout the summer.

Key to the Species
1. Plants light green to olive-green; erect axes with branchlets in two rows, forming triangular to lanceolate fronds . . . . *B. plumosa*
1. Plants dark green; erect axes with branchlets in two rows, all except those at tips of axes approximately the same length to form linear-lanceolate-shaped fronds . . . .*B. pennata*

B. pennata. These dark green plants consist of pinnately branched, coenocytic tubes up to 4 inches (10 cm) long. They are attached by rhizoids usually growing in tuftlike mats. The thallus contains many nuclei and discoid chloroplasts, which contain pyrenoids. The lateral branchlets are in two opposite rows that form a featherlike frond. The lateral branchlets may become cut off from the main axis and function as gametangia. The species is collected in protected areas in shallow water during the warmer part of the year.

B. plumosa. The plant forms dark green tufts up to 3 inches (7 cm) long. The coenocytic thallus is upright with main axes that are simple or branched. The thallus is variable in form but is usually lanceolate and up to ¼ inch (5 mm) wide. The fronds are usually triangular-shaped as the branchlets progressively decrease in length toward the apex. The species is collected in protected areas in shallow water during the warmer part of the year.

BRYOPSIDACEAE

▲ *Bryopsis plumosa* branching (close-up).

▲ *Bryopsis plumosa* branching (pressed specimen).

▲ *Bryopsis pennata* branching (pressed specimen).

◄ *Bryopsis pennata* (close-up).

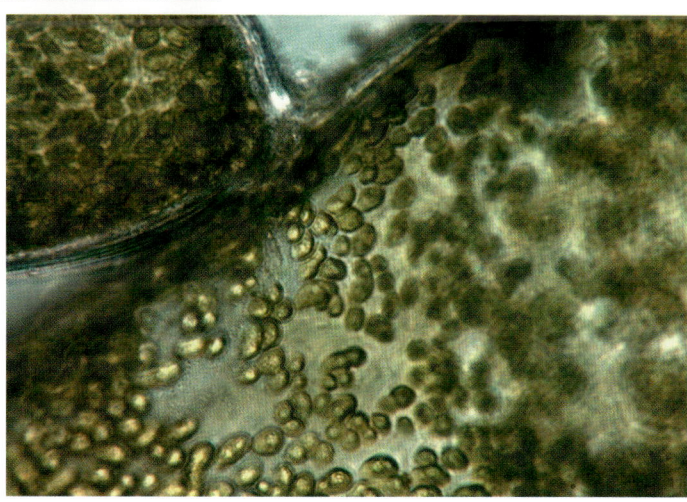

Bryopsis pennata discoid chloroplasts.

Family CODIACEAE
CODIUM
Codium taylorii P. C. Silva
⇒ KOH-dee-um tay-LOR-ee-eye

The thallus of this plant is upright and arranged into compact hemispherical clumps, up to 6 inches (15 cm) high. It is dark green, soft with a smooth surface, but covered with fine hairs and branching mostly dichotomously. The thallus has cylindrical to club-shaped utricles with round to slightly pointed apices from the sides of which emerge reproductive structures (gametangia). The holdfast has a crustlike base. The plants are found on jetty rocks in areas near channels and other passes during the summer.

Codium taylorii in situ.

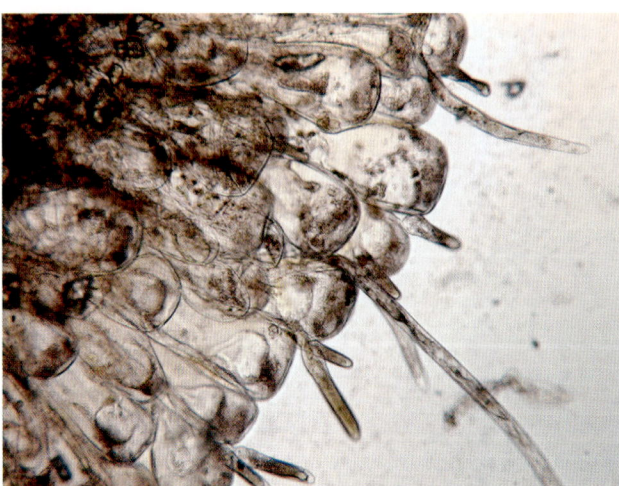

Codium taylorii utricles with gametangia and hairs.

Family CAULERPACEAE
CAULERPA
  ⟹ KAWL-er-puh
C. mexicana Sond. ex Kütz. [*C. crassifolia, C. c.* f. *mexicana*]
  ⟹ meks-sih-KAY-nuh
C. prolifera (Forssk.) J. V. Lamour.
  ⟹ pro-LEEF-er-uh

Key to the Species
1. Plant fronds (assimilators) divided, with pointed branchlets . . . .
 . . . . *C. mexicana*
1. Plant fronds flat, smooth, not toothed or divided; apex not
 indented or notched . . . . *C. prolifera*

C. mexicana. The plant is of a stoloniferous construction with rootlike structures (rhizoids) attached to a hard substrate or embedded in coarse substrate with upward-growing fronds (assimilators) that resemble flattened feathers. The assimilators are flat blades with broad, flat, upturned, pointed, divided branchlets. The internal portion of the thallus has inwardly projecting cell wall material called trabeculae. The alga is a warm-water plant and is found during the summer in tropical regions.

C. prolifera. The plant construction is similar to that of *C. mexicana* with a difference in the morphology of the assimilator. The plant has grass-green to dark green, flat, undivided, elongate blades that are thin and leathery. The blades often give rise to secondary blades off the center portion of the main assimilator. The plant is commonly found in seagrass beds during the warmer times of the year and in the southern tropical region of the coast.

Caulerpa mexicana morphology (close-up).

▲ *Caulerpa mexicana* habit.

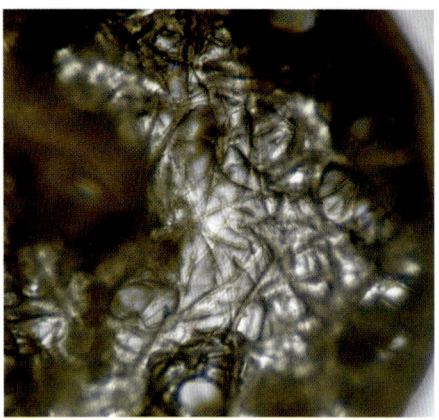

▲ *Caulerpa mexicana* morphology (pressed specimen).

▲ *Caulerpa mexicana* trabeculae.

◄ *Caulerpa prolifera* (pressed specimen).

Family UDOTEACEAE
HALIMEDA
Halimeda incrassata (J. Ellis) J. V. Lamouroux
▸ hal-ee-ME-da in-kras-SAY-tuh

The plants are light green, heavily calcified, hard, brittle, and distinctly ribbed with lobed segments connected by noncalcified flexible joints (geniculate). The segments branch initially in one plane. The lower portion of the stalk may be fused together and not flexible. The plant grows up to 10 inches (25 cm) tall. The species is located in the southern tropical areas of the Gulf Coast in seagrass beds attached to shallow coarse sediments such as shell hash or small rubble fragments.

Halimeda incrassata in situ in *Thalassia* seagrass bed.

Halimeda incrassata branching (close-up; pressed specimen).

PENICILLUS
Penicillus capitatus Lamarck
▶ pen-ih-SIL-us kap-ih-TAY-tus

The plant is stiff, calcified, and brushlike, up to 2–4 inches (5–10 cm) high, resembling a shaving brush, hence the common name Neptune's shaving brush. It has a faded green, well-defined, heavily calcified stipe with a rounded, lightly calcified cap. The cap is usually as long as broad and is composed of slender, dichotomously branched filaments. The plant is found widely scattered throughout seagrass beds in the tropical areas of the Gulf Coast.

Penicillus capitatus habit in seagrass bed.

Penicillus capitatus plant showing stipe (stalk) (pressed specimen).

Penicillus capitatus (close-up).

Order Dasycladales
Family DASYCLADACEAE
BATOPHORA
Batophora oerstedii J. Agardh
⇒ bat-oh-FOR-a or-STED-ee-eye

The plants grow as clusters of erect stipes with a series of horizontal branches arranged in whorls, producing a small, fuzzy, green to yellow-green cylinder. The erect portion is only about 1 inch (2–3 cm) in height and has a diameter of ¼ inch (5 mm). Reproductive structures appear as small green spheres on the divisions of horizontal branchlets. The spheres will ultimately develop into the reproductive bodies. Similar to *Acetabularia*, *Batophora* grows in protected, shallow areas of bays and estuaries attached to shell fragments and other hard substrates. It is most often found during the middle of the summer in the southern tropical areas of the Gulf Coast.

Batophora oerstedii (close-up).

Batophora oerstedii in situ.

Batophora oerstedii in *Thalassia* seagrass bed (close-up).

Family POLYPHYSACEAE
ACETABULARIA
⮕ a-si-tab-yoo-LAR-ee-uh
A. crenulata J. V. Lamouroux
⮕ kren-yoo-LAY-ta
A. farlowii Solms-Laubach
⮕ far-LOW-ee-eye

These plants are composed of a slender, calcified stipe 2½ inches (7 cm) tall bearing a concave disk at the top (sometimes as a series of superimposed disks). The disk is ½–¾ inch (6–18 mm) in diameter and composed of 30–80 moderately calcified segments (rays). A crown of projections (corona superior) arises from the center, and each segment bears a colorless hair. A similar structure (corona inferior) is located on the underside of the disk, but it lacks hairs. Many green spherical cysts (200–500) are borne within each of the rays. Plants generally form dense clusters of stipes with attached disks on quahog and oyster shells or other solid substrates in shallow bays along the coast. The two species are distinguished by the presence or absence of a short, central spine at the end of each ray. The plant is commonly called mermaid's wine cup.

Key to the Species
1. Disk ray with a short, central spine on outer margin . . . . *A. crenulata*
1. Disk ray with smooth, rounded outer margin . . . . *A. farlowii*

A. crenulata. The alga is sometimes heavily calcified and consists of an unbranched, white stipe, up to 1½ inches (4 cm) tall, which bears a concave, green disc, up to ⅜ inch (1 cm) in diameter. The vegetative thallus is uninucleate, diploid, and consists of a base of rhizoids. This erect axis bears a whorl of branched, hairlike appendages that often leave scars on being cast off. When reproduction begins, the nucleus divides into daughter nuclei that move up into the green disk, which is a whorl of developing gametangial rays at the apex of the stipe. Each ray has a single, central, short spine on the outer margin. There may be more than one disk formed in succession. These gametangia as mature rays will produce up to 500 spherical cysts per ray. This locally abundant alga is found attached to shells, rocks, and other hard substrates in protected bays during the warmer part of the year.

A. farlowii. The alga is found either solitary or in clusters and is heavily calcified. The thallus consists of an unbranched, white stipe, up to 1½ inches (3 cm) high, which bears a solitary, concave, bright green disk, up to ½ inch (12 mm) in diameter. The disk develops 20–30 (up to 40) rays that have a rounded outer margin, and each contains 40–120 spherical cysts. The alga is found attached to hard substrates on sandy bottoms, especially shell fragments in protected bays, and often grows intermingled with *A. crenulata*. It is found in abundance during the warmer times of the year.

▲ *Acetabularia crenulata* reproductive caps (close-up).

◀ *Acetabularia crenulata* cysts.

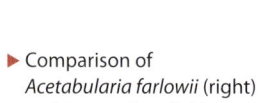

▶ Comparison of *Acetabularia farlowii* (right) and *A. crenulata* (left).

Acetabularia crenulata habit.

Acetabularia crenulata spines.

Appendix
Collection and Preparation Techniques

Important note: There are numerous laws that are relevant to the collection of specimens. In some cases, collecting may not be allowed or permits may be required for some locations. It is your responsibility to be aware of and adhere to these rules.

The purposes of taxonomic collections are to characterize the normal population in size and appearance and to provide the student with a sample for study. It is essential to collect representative samples (seasonally), being dutiful to collect the entire plant (including the holdfast). It may also be useful to collect samples of the same plant type from various habitats. Always maintain a field notebook and include information about the habitat, depth, site location, wave energy, and so on, as this will be essential for construction of the label, which may be as important as the specimen itself.

Collecting along the coast is simple and requires only a few pieces of equipment. A plastic bucket, zip plastic bags, putty knife for scraping, and protective clothing (water shoes, pants, and long-sleeved shirt) are all that is necessary for collecting from the shore or jetty. However, if you decide to collect in the water (snorkel or wading), a bathing suit may be all that is necessary, but remember it provides little or no protection against cuts, scrapes, and sunburn. A wetsuit may be used for protection or when the water is cold. Snorkeling equipment may aid in the collection of specimens below the water's surface.

If possible, avoid times when the wind speed is high and surf is crashing on the shore. Also, an incoming tide at passes (jetty) will usually bring less turbid oceanic water and aid in water clarity. But often the water is just not clear enough to see the seaweeds, and you must collect by feeling the plants along the jetty rocks below the surface. Be careful and move the hand/fingers gently to avoid injury by sharp edges, urchins, barnacles, and so on. You will soon learn that the algae have a characteristic texture or "feel," and you will be able to identify the specimens even before seeing them. If you choose to wear gloves for protection, you will lose the ability to identify the specimen by touch.

Other hazards of which you must be aware are the possibility of an "undertow" or other currents that may occur around the jetties; rogue waves when a boat or ship passes by; venomous animals (jellyfish, Portuguese man-of-war, stinging hydroids); sharp edges and punctures from oyster shells, jetty rock, urchins, cans, bottles, or fishing gear; and sun-induced skin burns along with possible dehydration.

In seagrass beds and shallow bay waters, collecting can be done by wading and snorkeling. Move slowly along the surface, looking through a glass-bottom box or while snorkeling, and examine shells, rocks, reef material, and seagrasses as well as sponges, tunicates, buoys, and submerged and floating wood.

It is best to sort collected specimens by taxa while still in the field. Individual plants are usually preserved in separate plastic Whirl-Pak or zip plastic bags with waterproof labels.

It is important that dry-pressed specimens be prepared carefully so that important morphological characters are displayed as fully and completely as possible. Portions of the specimens (fruiting structures or thallus sections) may be removed and placed in a vial of preservative for microscopic observation, which often is essential for identification. The following procedure, with a bit of practice, should produce good-quality dried specimens.

Rinse specimens free of any sand or debris. (Tap water may be used for this.) Remove any artifacts (shells, animals) that are not part of the specimen (although records of their presence should be made for ecological reference). If the holdfast is too thick and resistant to be pressed, either split it to remove portions or remove the holdfast and dry it separately (properly tagged for reference to the original specimen).

Preparation of Herbarium Mounts: Angiosperms and Seaweeds

Materials

    Plant press
    Blotters
    Newspaper
    Herbarium sheets
    Wax paper
    Corrugated ventilators
    Bulb syringe (pipette)
    Scissors (small)
    Nylon cloth
    Paintbrush (small)
    Paper towels
    Shallow pan

Angiosperm Procedure: Angiosperms, Halophytes, Seagrasses

1. Fold a single sheet of newspaper in half. On either the outside long edge or short edge, write the collection number and the date of collection.

2. Remove the plant from your collecting bag, leaving the rest of the plants covered so that they will not wilt. Wash any remaining dirt off the roots, and spread the plant out in a little less than the space of a half newspaper sheet. It is important to take plenty of time: *Don't hurry!*

3. Spread each leaf out flat. Turn over at least two leaves to show the underside, as the hairiness of that underside is often a crucial part of identification. Carefully spread open the flowers, making sure every petal lies smoothly. Try to press the flower in such a way as to best show important characteristics. Try to show the interior of at least one flower even if the important characteristics are unknown to you. Some of the flowers should lie with the lower side uppermost, as did the leaves.

4. If the plant is too long for the sheet, fold it sharply in a "V"; if that is not enough, fold it into an "N"; and if that is not enough, fold it into a "W." If it is still too big, cut it and use only part or press it in several sections. Be sure to describe the original size of the plant in your field notebook.

5. Small strips of wet newspaper are effective in holding down thinner flowers and leaves and do not have to be removed before pressing.

6. Open the plant press, lay a corrugated cardboard ventilator on one grid, then a heavy thickness of newspaper on that, and then add the pressed plant on the paper. If heavy newspaper is used as a blotter, slide in the folded single sheet of newspaper, with plant inside, so that its open edge is against the closed edge of the thick folded newspaper. This helps hold everything in and makes changing the heavier paper easier.

7. When all the plants have been pressed, place them, as described, in a series of sandwiches in the plant press and strap the press shut, as tightly as possible. Having one student stand on the press while the other pulls the straps tight helps produce flat specimens. Loose strapping gives wrinkled leaves and flowers.

8. Set the press where a current of warm, not hot, air can flow up through the ventilators. A large drying cabinet with light/heating element and a fan may be used.

9. Most plants dry in two or three days to a week. After 24 hours, go through and change or remove the blotters/newspaper. This is very important. The faster the plants dry, without high heat, the better the flower color will be retained. After removing the blotters, return the press, strapped shut again, to the warm drying area.

Seaweeds Procedure: Algae, Seaweeds

1. Because these plants collapse when removed from the water, in order to press them, first float them out in a shallow pan of water (saltwater if they are marine plants). Slide a piece of herbarium paper underneath the floating plant; use a piece only large enough to take the entire plant. Arrange the delicate thalli and branches carefully with the fingers, a narrow brush, or blunt probe or use water from a pipette to aid in spreading the fine branches and structures. Slowly withdraw the paper from the water, tilting it slightly to allow the water to flow off gently.

2. Drying algae in a press is similar to drying land plants, except that the top of the plants must be covered with something to prevent their sticking to the enclosing paper or blotter. A piece of white nylon lining material purchased from a fabric store is best. It is cheap, reusable, and dries very quickly. Write the collection number and date on the outside margin of the herbarium sheet. Care must be taken to ensure that in every step in the process accurate records are maintained. Some marine algae grow to great size, in which case only part of the plant can be pressed while the rest must be described. Some become very brittle and do not adhere well after drying. These can be carefully removed from the herbarium sheet (after drying) with a scalpel or single-edged razor blade and reglued with white glue (Elmer's) to a clean sheet of herbarium paper.

3. Algae are also preserved in fluids such as 3%–5% formalin in seawater, 70% ethyl alcohol, or FAA (formalin-acetic alcohol). Transfer the algae to new solution and into an appropriate-sized jar or vial upon returning to the laboratory. Again, do not forget a permanent identification tag inside the container and a label on the outside. If the containers will remain on the shelf for a long period of time, it is often necessary to dip the top of the container, cap and all, into melted paraffin (wax) and allow to cool. This will help prevent evaporation of the preservative.

4. To complete the documentation, prepare an adequate label to accompany the field or collecting number that was first assigned to the specimen and add necessary cross-indexing to the herbarium materials. Any preserved material or slides containing a subset of the material, for example, should be noted on the herbarium sheet.

5. A voucher specimen is fully prepared when
   a. it is a dried, pressed, herbarium specimen that is adequately labeled,
   b. it is a dried or wet-preserved specimen that is labeled and in a container, or
   c. it is a small specimen that is mounted on a prepared microscope slide and labeled.

A label should contain the following information:
- Family
- Genus, specific epithet, and author of the species name
- Common name (if it has one)
- Locality (country, state, county, etc.)
- Habitat notes
- Miscellaneous observations
- Collector's full name
- Collection number
- Date
- Name of the person identifying the plant

Preparation of Seaweed Permanent Microscope Slides

Materials
Plant material (preserved)
Cytoseal 60 or Karo syrup
Single-edged razor blade
Stains (if necessary)
Microscope slides
Toluene or xylene
Bulb syringe (pipette)
Scissors (small)
Cover slides
Alcohol (70%)
Paper towels
Labels

Slides Procedure

1. Voucher specimens usually require the preparation of microscopic samples of algal (seaweed) material. This is especially important in the collection of reproductive and other microscopic structures from field and pressed specimens that are necessary for the plant's identification. Microscope slides and slips may be purchased at hobby shops.

2. The specimen to be mounted must have been preserved earlier in the field or laboratory. Place the alga or structure from an alga on a microscope slide, and arrange it so that it is in one plane. Add a drop of 0.5%–1.0% aqueous stain and fix (if necessary, usually not). Rinse the stain off by flooding with distilled water and holding the slide at a slight angle over a paper towel to drain. The excess stain can also be carefully removed by blotting along the edges of the thallus with a piece of paper towel or bibulous paper.

3. Add a drop of alcohol, whisk it away with a blotter, and allow to air-dry. Repeat the procedure with a drop of toluene or xylene. Add a drop or two of Cytoseal 60 or Karo syrup to the specimen.

4. Carefully add a cover slip, one edge resting on a probe or forceps, and gently lower it over the mounting medium surface. Gently work any air bubbles out from under the cover slip. The slides must be kept on a flat surface. It may be helpful (especially in high humidity) to place the slides on a slide warming table.

5. When the slide is dry, add a proper label. Store the slide flat for 2–4 weeks, and finally store it on edge in a standard slide box.

Glossary of Selected Terms

This glossary provides definitions of many of the more common terms used in keys to various groups of marine algae and coastal plants. It is by no means exhaustive and includes some terms that are not used in this text but may be useful in further study.

abiotic. Not biotic; referring to the physical and chemical components of the environment.
acuminate. Pointed.
alga (algae; adj. = algal). A photosynthetic, nonvascular organism that reproduces without multicellular jacketed male and female structures.
alternate branching. Branching along two sides of the stalk, alternating at different levels or positions.
amorphous. Lacking distinctive morphology; without defined external structures.
anastomoses. Cross connections.
anatomy. The internal structure of an organism.
annular. Ringlike in structure; forming a ring, often as constrictions alternating with ridges.
anterior. In the frontal position; opposite of posterior.
antheridium (antheridia). Male reproductive structure that produces spermatia.
anticlinal. Radiating outward from the center.
apex. Distal end; outer tip; uppermost point; summit.
apical cell. The embryonic cell at the apex of a thallus.
apical depression. A pitlike cavity at the tip of a branch.
appendage. An outward extension.
articulate. Jointed.
asexual. Referring to a reproductive process unrelated to the fusion of genetic material.
assimilators. Erect, photosynthetic units of a thallus.
asymmetric. Showing a lack of symmetry; uneven.
attenuate. Tapering.
author(s). Person(s) who described a given species, whose name(s) appear(s) after the Latin generic and specific epithets; the last part of the complete scientific citation of an organism.
authority. Person with expertise (e.g., in a given taxon or group), not to be confused with the author(s) of a species.
auxiliary cell. In most red algae, a cell of the carpogonial filament or a related filament

that fuses with the carpogonium or an outgrowth of the carpogonium and may subsequently give rise to the gonimoblasts.

axial. Pertaining to the primary or central filament, the central core.

axil. Angle formed between a branch and the axis or main branch from which it arises.

axis (axes). Main stem or major branch; central line on which other parts are regularly arranged; the stemlike stalk.

basal. Toward the base or region of attachment.

basipetal. Referring to growth toward the base.

benthic. Attached to a definite substrate or anchored in sand or mud.

bilateral. Referring to a two-sided (opposite) arrangement in reference to a center line.

biogeography. The study of the geographical distribution of organisms.

biotic. Pertaining to living organisms.

bipinnate. Twice-branched, with both primary and secondary branches pinnately divided.

biseriate. In two rows.

bispore. One of two spores borne in pairs.

bladder. A hollow, inflated, gas-filled tissue that enables some algae to float or remain erect.

blade. Leaflike structure of an alga, often termed "frond"; leaf of a seagrass.

brackish water. Part seawater and part freshwater; diluted seawater.

branch. Main lateral appendage of the axis.

branchlet. Smaller projection off a main axis or larger branch.

bulbous. Shaped like a bulb; having a swollen aspect.

caespitose. In tufts.

calcareous. Encrusted with lime (calcium or magnesium carbonate).

calcification. Deposition of calcium carbonate within, between, or on cell walls.

calcified. Having lime deposits (calcium carbonate, a chalklike substance) within or on the plant; heavily calcified plants appear stonelike in texture.

capitate. Having a headlike end (with a necklike constriction).

carpogonial branch. Specialized internal filament terminating in a carpogonium.

carpogonium. The female sex organ in the red algae.

carposporophyte. A mass of tissue derived from a diploid zygote (fertilized carpogonium) and giving rise to carpospores.

cartilaginous. Hard and tough, often shiny.

central axis. The filament in the thallus of some red algae that gives rise to all other cells of the thallus; often visible in cross section as a single centrally located cell.

cervicorn. Referring to dichotomous branching with one arm of the dichotomy suppressed.

cf. Compared with; similar to but not quite the same as.

character. Feature used in taxonomy; one of the attributes that makes up or distinguishes an organism or individual.

chlorophyll. Green photosynthetic pigment.

chloroplast. Membrane-bound structure within the cell cytoplasm containing chlorophyll (green) pigments.

chromatophore. Pigmented structure within the cell cytoplasm.
class. Group of related organisms forming a category ranking above order and below phylum (name ending in "phyceae").
classification. Systematic arrangement in hierarchical groups or categories according to evolutionary (phylogenetic) relationships.
clathrate. Latticelike; perforated.
clavate. Club-shaped.
coenocytic. A type of thallus construction characterized by the absence of cross walls.
colony. An organism composed of interacting connected individuals.
community. Distinctive assemblage of organisms with one to several characteristic species.
compressed. Somewhat flattened; oval in transverse section.
concave. Rounded inward like the inside of a bowl; arched or curved inward; referring to a smoothly curved depression.
concentric. Having curved zones that are parallel.
conceptacle. A cavity (containing reproductive cells) with an opening to the outside.
conical. Shaped like a cone.
conspecific. Identical with another species.
constructed. Narrowed or contracted.
convergent. Approaching and finally merging together; coming together.
convex. Rounded outward; like the outside of a sphere; arched or curved outward; bulging outward.
coralline algae. Calcareous red algae of the order Corallinales.
cordate. Heart-shaped.
core. Central tissue or central filaments generally running parallel to the surface; often termed "medulla."
coriaceous. Leathery.
corona. Whorl of appendages above or below the center of the gametangial rays in *Acetabularia* and related genera.
cortex. The outer portion of a fleshy organ, assuming a central medulla is distinguishable.
cortical. Pertaining to the cortex.
corticated. Having cortication.
cortication. Filamentous or pseudoparenchymatous outgrowths that conjoin at least in part and form a new outer covering of the thallus or stipe.
corymbose. Having longer dichotomies at base, decreasing in length outward to create a flat-topped cluster.
cosmopolitan. Distributed throughout the world.
cruciate. Having contents of a tetrasporangium divided in two or three planes at right angles to one another.
crust. Two-dimensional layer prostrate on the substrate.
crustose. Crustlike.
cryptic. Concealed; difficult to distinguish.
cryptostoma (cryptostomata). Open cavity or pit on a thallus surface containing sterile hairs (paraphyses).

cuneate. Wedge-shaped.
cuticle. Structureless transparent layer on the outer surface of some plants.
cyst. Asexual resting-stage structure; thick-walled sphere of dense protoplasm that germinates to form a new individual.
cystocarp. A reproductive structure in the red algae, bearing carpospores or tetraspores or formed after fertilization of the carpogonium.
deciduous. Not permanent; falling off or shedding seasonally or at a certain stage in development.
decumbent. Lying flat, but with the summit ascending.
dentate. Toothed.
denuded. Left bare; loss of lateral filaments or branchlets.
determinate. Having limited growth.
diameter. A simple comparative measurement of the ratio of the width of a cell compared to the cell's length. Example: Length of cell = 3 widths or diameters.
dichotomous. Branched by repeated forks.
dioecious. Having distinct male and female plants.
discoid. Disklike; flat and circular.
discrete. Separate; not joined or coalescent.
distal. Referring to the farthest point; opposite of basal or proximal.
distichous. In two vertical ranks or rows.
ditrichotomous. Branching both in a dichotomous and a trichotomous manner into two equivalent parts.
divaricate. Extremely divergent.
dominant. Referring to the most abundant, most conspicuous organism; having the greatest ecological influence.
dormant. Inactive; without active growth.
dorsal. Referring to the upper surface; opposite of ventral.
dorsiventral. Having distinct dorsal and ventral surfaces.
drift. Unattached macrophytes suspended in the water or lying loosely on the bottom; when abundant, often driven into heaps; to be moved or carried along by currents.
ecology. The study of organisms in relation to their environment.
ecortication. Without cortication.
ecosystem. A system composed of interacting biotic and abiotic elements functioning as a unit.
ellipsoid. Forming an ellipse; symmetrical at both ends.
encrusting. Crustlike; covering or overlaying with a thin layer.
endogenous. Arising from internal tissues.
entire. Having a smooth margin; not toothed or lobed.
entwined. Tangled; twisted together; intermixed.
epidermis. The surface layer of cells.
epilithic. Attached to rock or stone.
epiphyte. An organism that lives upon, but does not get its nourishment from, a plant.
et al. And others; used in citations of authors' names to indicate subsequent authors.

eucaryotic. Possessing membrane-bound organelles such as a nucleus and chloroplasts.
eutrophic. Rich in dissolved nutrients.
eutrophication. Becoming eutrophic and often deficient in oxygen; nutrification; becoming rich in dissolved nutrients.
ex. From or according to; used in author citations to connect two authors, with the second validating a name proposed by the first.
exogenous. Arising from surface tissues.
falcate. Sickle-shaped.
family. Group of closely related organisms ranking above genus and below order (name ending in "aceae").
fasciculate. In tufts.
fibrous. Containing or consisting of fibers; capable of being divided into slender strands or threads.
filament. Very slender, threadlike, single series of cells; referring to any very slender thread, fiber, or strand; a group of cells attached end to end.
filiform. Threadlike.
flabellate. Fan-shaped.
flaccid. Limp; flabby.
foliose. Leaflike; broadly flattened.
fountain type. Having multiaxial construction with numerous axial filaments.
form (abbreviated f.). Taxonomic subdivision of a species indicating a variation resulting from environmental factors; the external shape of an organism; the morphology.
frond. Blade or leaflike structure of an alga.
fusiform. Spindle-shaped.
fusion. Adhesion of cells following contact.
gametangial thallus. Haploid thallus producing gametes by mitosis; the sexual phase or gametophytic phase in the life history.
gametangium (gametangia). Reproductive structure producing gametes (in Rhodophyta, female gametangium = carpogonium; male gametangium = spermatangium).
gamete. Haploid (containing a single set of chromosomes, or N) reproductive cell capable of uniting with another gamete to form a diploid (two sets of chromosomes, or 2N) zygote.
gametophytic phase. Haploid or gametangial thallus producing gametes by mitosis; the sexual phase in the life history.
gelatinous. Slippery and mucilaginous.
genus (genera). Group of closely related organisms consisting of one or more species; first Latin word of a scientific name.
gland cell. A secretory cell, often appearing opalescent.
gonimoblast. A short filament from the auxiliary cell or the carpogonium base, usually curving upward and terminating in a carposporangium.
habit. The overall morphological form or outward appearance of an organism.

habitat. The physical environment in which an organism lives; the place where an organism is typically found.

hair. Elongated, finely filamentous, threadlike, uniseriate strand or a single colorless cell, often deciduous, borne on outer cortex.

hair cell. Basal cell from which a hair is attached to the cortex or surface; also termed trichocyst or megacell.

hapteroid. Containing lobed, branched, or fingerlike extensions of a cell that can attach to adjacent structures.

hapteron. An attachment organ; in many red algae a unicellular or multicellular organ in which the tip is markedly flattened where it is in contact with the substrate.

herbarium (herbaria). An organized collection of plants.

herbivory. The consumption of plant or algal material.

heterocyst. A cell that is conspicuously larger and often thicker walled than adjacent vegetative cells.

holdfast. A basal structure or cell specialized as an organ of attachment.

holocarpic. Refers to the conversion of all the protoplasm into gametangia; upon release of reproductive structures, the thallus disintegrates.

holotype. The single (type) specimen to which the name of a taxon is permanently attached.

homonym. An invalid name because the same name was used earlier for a different plant (name incorrectly attached to a different type, specimen).

hyaline. Clear and colorless.

hypothallus. Lowermost layer or zone of a prostrate thallus.

incurved. Curved inward toward the main axis.

indeterminate. Having more or less unlimited growth.

indusium. Membrane that covers the surface of a reproductive structure.

inferior. Below some other structure.

inrolled. Rolled inward toward the central axis or toward the upper surface.

intercalary. Within; between two cells or cell layers.

internode. As applied to algae, the interval on a (stemlike) thallus between two joints or "nodes."

intertidal. Pertaining to the area of the shoreline between the uppermost and lowermost tidal levels.

involucral cell. Large incurved, sausage-shaped cell encapsulating or protecting the reproductive structure.

involucral filament. One of the sterile filaments surrounding or enveloping the reproductive structure.

iridescent. Having a surface sheen reflecting an interplay of neon/metallic colors; glowing; shining.

irregular. Without a defined branching pattern.

joint. Junction between segments; end walls between contiguous cells in a filament; often termed "node."

keeled. Having a narrow, lengthwise, prominent ridge, like the keel of a boat.

lamellate. Made up of thin plates or sheets.

lanceolate. Shaped like a lance point.

lance-shaped. Much longer than broad; tapering toward a pointed apex.
lateral. At, from, or toward the edges of a structure.
lax. Loose; limp; not rigid.
lectotype. Specimen selected as the nomenclatural basis of the taxon when the holotype is lost or destroyed.
life history. An organism's vegetative and reproductive phases.
linear. Narrow; ribbonlike.
locule. Compartment or chamber.
longitudinal. Lengthwise; in the direction of the longest axis; right angles to transverse.
loose lying. Not attached; lying unattached on the substrate.
macroalga. Larger alga easily observed without the aid of a microscope.
macrophyte. An aquatic plant easily observed without the aid of a microscope (includes both macroalgae and seagrasses).
macroscopic. Large enough to be recognized by the unaided eye; observable without a microscope.
medulla. Central core area of a massive thallus or stipe.
medullary cells. Differentiated inner cells, generally surrounded by a cortex; the central tissue.
meristematic. Referring to tissue or region of active cell division and growth.
microscopic. Unable to be reliably observed without the aid of a microscope.
midrib. A raised, usually cylindrical, "nerve" of the thallus surface.
moniliform. Barrel-shaped or beadlike.
monoecious. Having both sexes present in same individual.
monograph. Systematic account of a particular group at the generic level or above; to produce a monograph.
monosiphonous. Uniseriate.
monosporangium. Structure producing a single spore (monospore).
monospore. An asexual unicellular spore.
morphology. The external form of the thallus or structure.
mucilaginous. Slippery, slimy, or jellylike.
mucronate. Possessing a short, straight point.
multiaxial. Referring to growth from a group or core of central filaments.
multicellular. Composed of many cells.
multinucleate. Having more than one nucleus per cell.
multiseriate. In a number of rows.
node. The point from which a branch arises; the junction of two cells; a joint between two levels or segments of the thallus.
nomenclature. The system of naming; the use of rules to correctly determine the scientific name of an organism.
nutrient. Enrichment or chemical fertilizer essential for growth.
nutrification. Addition of nutrients.
oogonium (oogonia). Female reproductive organ producing one or more eggs (oospores).
oospore. A thick-walled zygote with food reserves; a fertilized female gamete.

opposite. Branching in which lateral branches regularly arise from a node in pairs.
orbicular. Like a disk; circular.
order. Category of taxonomic classification ranking above family and below class (name ending in "ales").
organelle. Small membrane-bound body within cytoplasm of cells serving specific functions.
oval. Having the form of an egg; shaped like the longitudinal transverse section of an egg.
ovate. Having the outline of a longitudinal section of an egg (two-dimensional).
ovoid. Shaped like an egg (three-dimensional).
palmate. Extending fingerlike or radiating fanlike from a point.
papillate. Covered with wartlike projections.
paraphyses. Sterile hairs.
parenchymatous. Referring to tissue resulting from cell divisions in more than one plane.
parent cell. Initial source or origin of subsequent cells or structures.
parietal. Appressed to the outer wall or toward the outside.
partition. Cross wall or cross membrane dividing a siphon or tube.
pelagic. Floating in offshore oceanic waters.
peltate. Parasol-shaped; flattened horizontal disk with a perpendicular stalk attached near center.
perennial. Persistent for several years.
pericarp. The outer wall of an enclosed cystocarp.
pericentral cells. Cells arranged in regular tiers about the central axial cell in certain red algae.
periclinal. Parallel and interior to the surface; longitudinal.
peripheral. At the outer perimeter; on the circumference.
perithallus. Middle layers or zone of a prostrate thallus.
phaeophycean hairs. Surface hairlike filaments with the growing region near the base (only present in the Phaeophyceae).
phycology. The study of algae; equivalent to and replacing the outdated term "algology."
phytoplankton. Minute photosynthetic organisms suspended in the water column.
pigmented. Having color or pigment.
pincerlike. Two-pronged, grasping, or clawlike.
pinnate. Having distichous branchlets arranged like the barbs of a feather.
pit connection. A cytoplasmic strand connecting two adjoining cells through a pit in their respective walls.
plane. Flat surface (e.g., growth in one plane results in a two-dimensional plant).
plasticity. The ability of an organism to change form in response to varying environmental conditions.
plastid. Pigmented photosynthetic organelle or body within the cytoplasm of a cell, generally containing chlorophyll and other pigments.
plumose. Featherlike.

plurilocular sporangium. A sporangium with numerous distinct, walled cells, with walls that persist even after the spores have been shed.
polychotomous. Branching many times from a terminal point or area.
polyhedral. Having many sides.
polysiphonous. In red algae, a type of construction in which the central axial cell is surrounded by three or more pericentral cells of similar length and arranged in regular tiers.
polysporangium (polysporangia). A sporangium containing many spores, generally more than four.
pore. An opening or hole, often at the apex of a reproductive structure.
posterior. In the rear position; opposite of anterior.
primary. Referring to the first part in development.
procarp. A unit group of reproductive cells in some red algae consisting of the carpogonium and the auxiliary cell as well as certain adjacent cells.
procaryotic. Without membrane-bound organelles as the nucleus, chloroplast, and so forth.
procumbent. Lying along the substrate.
proliferation. Thallus projection; branching structure.
proliferous. Reproducing easily by vegetative means; producing many branchlets.
propagate. To reproduce or spread.
propagule. A few-celled branch of definite form that can break loose from the parent plant and form a new plant by vegetative means.
prostrate. Horizontal in relation to the substrate.
protoplast. The contents of a cell.
proximal. Near point of attachment; lower part; basal; opposite of distal.
pseudo. False; looking similar but not exactly alike.
pseudodichotomous. Appearing dichotomous; formed by strongly developed lateral branch resembling the original axis.
pseudoparenchyma. Tissue resulting from a compact arrangement of filaments in which the identity of individual filaments is more or less obscured.
pulvinate. Cushionlike or cushion-shaped.
pyrenoids. Small, rounded structures within the chloroplast.
pyriform. Pear-shaped.
radial. Radiating away from the central point or axis.
ramuli. Branches.
receptacle. Modified branch or branchlet, usually enlarged or swollen.
recurved. Bent away from the main curvature of the axis where reproductive organs develop.
reniform. Kidney-shaped.
reticulate. Netted or covered with a layer, which is netlike.
rhizines. Slender, colorless filaments with embedded pyrenoids.
rhizoid. A slender, descending filament.
rhizoidal. Pertaining to the rhizoid.
rhizomatous. Having rhizomes, runners, or stolons.

rhizome. An underground, horizontal, stemlike structure similar to, but not the same as, stolons, which travel along the surface.
rigid. Stiff or inflexible.
root. The absorptive and anchoring organ of vascular plants (seagrasses).
ruffled. With a strongly waved or undulating margin.
rugose. Roughened or wrinkled.
runner. Stolon; rhizome; horizontal, stemlike structure.
salinity. Salt content of water, usually expressed in parts per thousand (ppt).
scale. Adherent leaflike structure differing from the normal leaves; generally small, thin, and nonpigmented.
scar cell. A small cell at the base of the thallus.
seagrass. Marine vascular plant growing submersed, at least at high tide.
seaweed. Larger marine plant (macrophyte) easily observed with the unaided eye.
secund. Arranged in a single plane.
sediment. Particulate matter that settles and accumulates on the bottom of bodies of water, includes deposits of mud, sand, and gravel.
segment. One central cell and surrounding pericentral cells (if present) and cortex (if present); segments are generally connected by joints (sometimes termed "nodes"); in calcified groups, referring to large, hard, calcified sections between short uncalcified flexible joints; often termed "internode."
sensu. In the sense (opinion, perception) of; as described by.
septum. A partition or wall trichoblast where it adjoins.
seriate. Arranged in a series.
series. Repetitive sequences.
serpulid reef. Hard substrate composed of the calcareous external tubes of marine segmented worms (polychaetes) found in Baffin Bay, Texas.
serrate. Marginally toothed; with pointed teeth.
sessile. Not stalked; attached directly to substrate or parent structure; nonmotile or attached.
seta. A hair or bristle.
sheath. The surrounding material; in vascular plants, the leaflike basal structure surrounding the stem.
simple. Unbranched.
sinuate. With a deep, wavy margin.
siphon. A more or less tubular structure without cell cross walls resembling a filament.
siphonaceous. Composed of tubes or siphons, generally multinucleate.
sorus (sori). A cluster, usually on the thallus surface of sporangia or other reproductive bodies.
specific epithet. The second Latin word in the scientific name, given to a group (i.e., species) of closely related organisms that can interbreed to produce fertile offspring.
spermatangial plant. Plant producing spermatia; male plant.
spermatium. A nonmotile male cell of the red algae.
spindle-shaped. Circular in transverse section; tapering toward each pointed end.

spine. Stiff, sharply pointed, elongated projection.
sporangium (sporangia). A spore-producing structure.
spore. Single-celled reproductive body that can germinate into a new thallus.
sporophytic phase. Diploid (2N) spore-producing stage in an algal life history.
spur. Short, hornlike projection.
stellate. Star-shaped.
sterile. Lacking reproductive structures; vegetative.
stichidium. A specialized branch bearing reproductive organs.
stipe. The stemlike or stalk portion of an alga.
stipitate. Having a stipe or stalk.
stolon. Creeping axis or surface runner connecting ascending fronds or branches and descending rhizoids (e.g., *Caulerpa* and certain other species of algae); horizontal or creeping stem, prostrate axis, runner with roots (descending) and shoots (ascending) at the nodes (e.g., seagrasses).
stoloniferous. Stolonlike; forming runners along the surface substrate from which the erect portions arise.
striated. Having fine, delicate lines or markings.
striations. Somewhat parallel, delicate, narrow markings.
stubby. Short, thick, and often rounded.
subapical. Immediately below the apex or tip.
subcortical. Immediately below the cortex.
sublittoral. The biological zone below the low-tide mark.
substrate. Substance or surface with which a benthic organism is associated.
subtidal. Below the lowest low-tide level.
superior. Above some other structure.
taxon (taxa). General term applied to any taxonomic entity irrespective of its level of hierarchical classification.
taxonomy. The discrimination of taxa or the general principles of scientific identification; the orderly recognition of organisms according to their presumed natural relationships.
tendril. Twining or clasping structure.
terete. Cylindrical; in cross section, circular.
terminal. Outer tip, distal end, apex, or end cell in a chain.
ternate. In threes.
tetrad. In fours.
tetrahedral. Having the contents of a tetrasporangium divided triangularly so that only three of the tetraspores can be seen at once.
tetrasporangium. A spore-producing organ whose contents divide into four parts.
tetraspore. A meiospore in red algae.
tetrasporic stage. The stage in which a diploid thallus produces haploid tetraspores by meiosis.
thallus. The entire plant body of an alga.
tiered. Layered; regularly stacked rows or series of cells arranged one above the other.
torulose. Beaded.

trabeculae. Simple or forked wall protuberances, often appearing as a meshwork.
transverse. Across or at right angles to the main axis.
transversely divided. Referring to division in a plane perpendicular to the long axis.
trichoblast. A simple or branched monosiphonous filament arising from or near the apex in the family Rhodomelaceae (Rhodophyta).
trichocyst. Specialized, hair-producing cell that is usually larger in size and thicker walled than other cells; often termed "megacell."
trichogyne. Threadlike upper portion of a carpogonium.
trichome. A hair (sometimes branched); in blue-green algae, the filament without its surrounding sheath.
trichothallic. A form of growth involving the division of intercalary cells in a group of terminal filaments, with subsequent adhesion of the basal parts of the filaments.
trichotomous. Branched by repeated forking into three equivalent parts.
trilobed. Having three lobes.
truncate. Cut off.
tubular. Tubelike; like a hollow cylinder.
tuft. Cluster of filaments; cluster of branches or siphons attached at a single basal area.
turf. Sparse to dense-growing, short, intertwined mat of small thalli (usually less than 1¼ inches [3 cm] high); layer of short upright shoots arising from prostrate bases.
type. The specimen to which a species (genus or other higher taxon) name is permanently attached.
understory. A group of organisms growing under larger organisms.
undulate. Wavy.
uniaxial. Having only one axial filament in the thallus.
unicellular. Single-celled.
unilateral branching. Having branches that arise only from one side of the stalk.
unilocular. Having only one cavity, that is, not divided up by numerous persistent cross walls as in a plurilocular structure.
uninucleate. Having only one nucleus per cell.
uniseriate. In a single row.
unistratose. Of one layer.
upright. Oriented vertically to the substrate.
urceolate. Shaped like a pitcher with a protruding mouth.
urn-shaped. Egg-shaped with slender protruding orifice; shaped like an amphora (water vessel).
utricle. An inflated end of a coenocytic filament; these are grouped to form an outer layer, as in *Codium*.
variety (abbreviated var.). Latin name added to the generic and specific epithets of a species to designate a consistent difference within that species but not substantial enough to separate as a new species.
vegetative. Referring to cells or tissue not associated with reproduction; nonsexual.
veinlet. Small, veinlike structure.
veins. Strands of larger cells within tissue; vascular-conducting tissue in seagrasses.

venation. Pattern formed by veins.
ventral. Lower or underside; opposite of dorsal.
verticillate. Branching in a flat whorl; radiating outward in all directions in a single plane.
vesicular. Saclike; commonly more or less spherical.
voucher specimen. Herbarium specimen kept as documentation for an identification reported in the literature.
whorled. Referring to branching in which three or more lateral branches regularly arise from a node.
wing. Thin, wide, lateral extension along an axis, midrib, branch, or any other structure.
zonate. Having contents of a tetrasporangium lying in a row as a result of division in three parallel planes.
zonation. Arrangement in horizontal and parallel zones or bands.
zygospore. Fertilized oospore (egg).

Bibliography

Albert, Erin M., and Roy L. Lehman. 2008. "Algal Community Structure of the East and West Flower Gardens, Northwestern Gulf of Mexico." *Texas Journal of Science* 60 (3): 201–214.

Arber, A. 1920. *Water Plants, a Study of Aquatic Angiosperms.* Cambridge: Cambridge University Press. 436 pp.

Dawes, C. J. 1981. *Marine Botany.* New York: John Wiley & Sons. 628 pp.

———. 1987. "The Dynamic Seagrasses of the Gulf of Mexico and Florida Coast." In *Proceedings of the Symposium on Subtropical-Tropical Seagrasses of the Southeastern United States,* edited by M. J. Durako, R. C. Phillips, and R. R. Lewis, 25–38. St. Petersburg: Florida State Department of Natural Resources.

den Hartog, C. 1967. "The Structural Aspect in the Ecology of Sea-Grass Communities." *Helgolander Wissenschaftliche Meeresuntersuchungen* 15:648–659.

———. 1970. *The Seagrasses of the World.* Amsterdam: North Holland Publishing. 275 pp.

———. 1979. "Seagrasses and Seagrass Ecosystems, an Appraisal of the Research Approach." *Aquatic Botany* 7:105–117.

Dunton, K. H. 1990. "Production Ecology of *Ruppia maritima* L. s.l. and *Halodule wrightii* Aschers. in Two Subtropical Estuaries." *Journal of Experimental Marine Biology and Ecology* 143:147–164.

Edwards, Peter. 1976. *Illustrated Guide to the Seaweeds and Sea Grasses in the Vicinity of Port Aransas, Texas.* Austin: University of Texas Press. 128 pp.

Eleuterius, L. N. 1990. *Tidal Marsh Plants.* Gretna, LA: Pelican Publishing. 168 pp.

Fassett, N. C. 1957. *A Manual of Aquatic Plants.* Madison: University of Wisconsin Press. 405 pp.

Fikes, Ryan, and Roy L. Lehman. 2008a. "The Occurrence of *Agardhiella ramosissima* (Gigartinales) and *Acanthophora spicifera* (Ceramiales) in the Texas Coastal Bend." *Texas Journal of Science* 60 (3): 221–224.

———. 2008b. "Small-Scale Recruitment of Flora to a Newly Developed Tidal Inlet in the Northwest Gulf of Mexico." *Gulf of Mexico Science* 26 (2): 130–132.

———. 2010. "Recruitment and Colonization of Macroalgae to a Newly Developed Rocky Intertidal Habitat in the Northwest Gulf of Mexico." *Gulf and Caribbean Science* 22:9–20.

Fikes, Ryan L., Roy L. Lehman, and Kyle V. Klootwyk. 2010. "An Update on the Benthic Algae of Mansfield Pass, Texas." *Texas Journal of Science* 62 (3): 183–194.

Fonseca, M. S. 1994. *Seagrasses: A Guide to Planting Seagrasses in the Gulf of Mexico.*

Texas A&M University Sea Grant College Program. Technical Report TAMU-SG-94-601. 26 pp.

Godfrey, R. K., and J. W. Wooten. 1981. *Aquatic and Wetland Plants of Southeastern United States: Dicotyledons.* Athens: University of Georgia Press. 933 pp.

Gould, F. W. 1965. *Grasses of the Texas Coastal Bend.* College Station: Texas A&M University Press. 189 pp.

Jewett-Smith, J. 1991. "Factors Influencing the Standing Crop of Diatom Epiphytes of the Seagrass *Halodule wrightii* Aschers. in South Texas Seagrass Beds." *Contributions in Marine Science* 32:27–38.

Jones, F. B. 1982. *Flora of the Texas Coastal Bend.* Corpus Christi, Tex.: Mission Press. 267 pp.

Jones, S. D., J. K. Wipff, and P. M. Montgomery. 1997. *Vascular Plants of Texas.* Austin: University of Texas Press. 404 pp.

Lehman, Roy L. 1999. "A Checklist of Benthic Marine Macroalgae from the Corpus Christi Bay Area." *Texas Journal of Science* 51 (3): 241–252.

———. 2007. "Reef Algae." In *Coral Reefs of the Southern Gulf of Mexico,* edited by J. W. Tunnell Jr., Ernesto A. Chavez, and K. Withers, 87–94. College Station: Texas A&M University Press. 194 pp.

Lehman, Roy L., Ruth O'Brien, and Tammy White. 2005. *Plants of the Texas Coastal Bend.* College Station: Texas A&M University Press. 352 pp.

Littler, D. S., M. M. Littler, and M. D. Hanisak. 2008. *Submersed Plants of the Indian River Lagoon.* Washington, D.C.: Offshore Graphics. 286 pp.

Loraamm, L. P. 1980. "Flowers That Took to the Sea." *Sea Frontiers* 26:84–90.

McCoy, E. D., and K. L. Heck. 1976. "Biogeography of Corals, Seagrasses and Mangroves: An Alternative to the Center of Origin Concept." *Systematic Zoology* 25:201–210.

McMahan, C. A. 1968. "Biomass and Salinity Tolerance of Shoal Grass and Manatee Grass in the Lower Laguna Madre, Texas." *Journal of Wildlife Management* 32:501–506.

McMillan, C. 1976. "Experimental Studies on Flowering and Reproduction in Seagrasses." *Aquatic Botany* 2:87–92.

———. 1985. "The Seed Reserve for *Halodule wrightii, Syringodium filiforme* and *Ruppia maritima* in Laguna Madre, Texas." *Contributions in Marine Science* 23:141–149.

McMillan, C., and F. N. Moseley. 1967. "Salinity Tolerances of Five Marine Spermatophytes of Redfish Bay, Texas." *Ecology* 48:503–506.

Merkord, G. W. 1978. "The Distribution and Abundance of Seagrasses in Laguna Madre of Texas." Master's thesis, Texas A&I University. 56 pp.

Moffler, M. D., and M. J. Durako. 1980. "Observations on the Reproductive Ecology of *Thalassia testudinum* (Hydrocharitaceae) in Tampa Bay." *Florida Scientist* 43 (Suppl. 1): 8.

Morgan, D., and C. L. Kitting. 1984. "Productivity and Utilization of the Seagrass *Halodule wrightii* and Its Attached Epiphytes." *Limnology and Oceanography* 29:1066–1076.

Phillips, R. C. 1960. "Observations on the Ecology and Distribution of the Florida Sea Grasses." *Professional Paper Series, Florida Board of Conservation* 2:1–72.

———. 1978. "Seagrasses and the Coastal Marine Environment." *Oceanus* 21 (3): 30–40.

Phillips, R. C., and E. G. Meñez. 1988. *Seagrasses.* Smithsonian Contributions to the Marine Sciences No. 34. Washington, D.C.: Smithsonian Institution Press. 104 pp.

Richardson, Alfred. 2002. *Wildflowers and Other Plants of Texas Beaches and Islands.* Austin: University of Texas Press. 247 pp.

Schneider, C. W., and R. B. Searles. 1991. *Seaweeds of the Southeastern United States.* Durham, N.C.: Duke University Press. 554 pp.

Silberhorn, G. M. 1982. *Common Plants of the Mid-Atlantic Coast: A Field Guide.* Baltimore: Johns Hopkins University Press. 256 pp.

Sorensen, L. O. 1979. *A Guide to the Seaweeds of South Padre Island, Texas.* Scottsdale, Ariz.: Gorsuch Scarisbrick Publishing. 123 pp.

Stutzenbaker, C. D. 1999. *Aquatic and Wetland Plants of the Western Gulf of Mexico.* Austin: Texas Parks and Wildlife Press. 465 pp.

Taylor, W. R. 1960. *Marine Algae of the Eastern Tropical and Subtropical Coasts of the Americas.* Ann Arbor: University of Michigan Press. 870 pp.

Thayer, G. W., S. M. Adams, and M. W. LaCroix. 1975. "Structural and Functional Aspects of a Recently Established *Zostera* Marina Community." In *Estuarine Research,* vol. 1, edited by L. E. Cronin, 517–540. New York: Academic Press. 587 pp.

Tiner, R. W., Jr. 1987. *A Field Guide to Coastal Wetland Plants of the Northeastern United States.* Amherst: University of Massachusetts Press. 286 pp.

Walters, D. R., and D. J. Keil. 1996. *Vascular Plant Taxonomy.* Dubuque, Iowa: Kendall/Hunt Publishing. 608 pp.

Woelkerling, W. J. 1976. *South Florida Benthic Marine Algae: Keys and Comments. Sedimenta V.* Miami Beach, Fla.: Comparative Sedimentology Laboratory, Fisher Island Station. 145 pp.

Wynne, M. J. 1986. "A Checklist of Benthic Marine Algae of the Tropical and Subtropical Western Atlantic." *Canadian Journal of Botany* 64 (10): 2239–2281.

———. 1993. "*Prionitis pterocladina* sp. Nov. (Halymeniaceae, Rhodophyta), a Newly Recognized Alga in the Western Gulf of Mexico." *Botanica Marina* 36:535–543.

———. 1998. "A Checklist of Benthic Marine Algae of the Tropical and Subtropical Western Atlantic: First Revision." *Nova Hedwigia* 116:1–155.

———. 2009. "A Checklist of Benthic Marine Algae of the Coast of Texas." *Gulf of Mexico Science* 26 (1): 64–87.

Zieman, J. C. 1982. *The Ecology of the Seagrasses of South Florida: A Community Profile.* FWS/OBS-82/25. Washington, D.C.: US Fish and Wildlife Service, Office of Biological Services. 158 pp.

Zimmerman, R. J., and A. H. Chaney. 1969. "Salinity Decrease as an Affector of Molluscan Density Levels in a Turtle Grass (*Thalassia testudinum* König) Bed in Redfish Bay, Texas." *Texas A&I University Studies* 2 (1): 5–10.

Index

Acanthophora spicifera, 91, 106–107
Acetabularia crenulata, 87, 155, 175–177
Acetabularia farlowii, 87, 155, 175–177
Acinetosporaceae, 138, 146
Agardhiella subulata, 92, 126–127
Aglaothamnion cordatum, 91, 100–101
Aglaothamnion halliae, 91, 100–101
Aizoaceae, 11, 13, 14
annual glasswort, 9, 29–30
Apiaceae, 11, 13, 16
Asteraceae, 11, 13, 18
Avicennia germinans, 75–77
Avicenniaceae, 75–77

Bangia fuscopurpurea, 91, 95–96
Bangiaceae, 91
Bangiales, 91, 95
Bangiophyceae, 91, 95
Bataceae, 11, 13, 20
Batis maritima, 9, 11, 20–23, 49
Batophora oerstedii, 87, 155, 173–174
beach evening primrose, 12, 37
black mangrove, 75–77
bluebell gentian, 12, 35–36
blue-green algae, 88
Bolboschoenus robustus, 12, 44
Boraginaceae, 11, 13, 23
Borrichia frutescens, 9, 11, 18
Botryocladia occidentalis, 93, 134
Brassicaceae, 11, 13, 25
brown seaweeds (algae), 88
Bryocladia cuspidata, 91, 108–109
Bryocladia thyrsigera, 91, 108–109
Bryopsidaceae, 154, 166
Bryopsidales, 154, 166
Bryopsis pennata, 154, 166–167
Bryopsis plumosa, 154, 166–167
buttonwood, 75, 78

Cactaceae, 3, 11, 13, 27
Cactus Family, 11, 27
Cakile geniculata, 11, 25–26
Cakile lanceolata, 11, 25
Calylophus serrulatus, 12, 36
camphor daisy, 9, 11, 19
Carolina wolfberry, 2, 3, 9, 12, 41
Carpetweed Family, 11, 14

Carrot Family, 11, 16
Cattail Family, 55
cattail, 10, 55
Caulerpa mexicana, 86, 154, 169–170
Caulerpa prolifera, 86, 154, 169–170
Caulerpaceae, 154, 169
Centroceras clavulatum, 91, 102
Ceramiaceae, 91, 100, 102
Ceramiales, 91, 100
Ceramium c. f. *flaccidum*, 91, 103–104
Ceramium cimbricum, 91, 103–104
Ceramium deslongchampii, 91, 103–104
Ceramium subtile, 91, 103–104
Chaetomorpha brachygona, 154, 160–161
Chaetomorpha gracilis, 154, 160–162
Chaetomorpha linum, 154, 160–162
Champia parvula, 93, 136
Champiaceae, 136
Chenopodiaceae, 11, 13, 28,
Chlorophyta, 88, 154, 156
Chondria capillaris, 91, 109–110, 115
Chondria cincophylla, 91, 109–110
Chondria dasyphylla, 91, 109–111
Chondria littoralis, 91, 109–111
Chordariaceae, 139, 148
Cladophora albida, 154, 163, 165
Cladophora dalmatica, 154, 163–164
Cladophora ruchingeri, 154, 163–164
Cladophora vagabunda, 154, 163–165
Cladophoraceae, 154, 160
Cladophorales, 154, 160
Cladosiphon occidentalis, 139, 148
clover grass, 60, 65–66
Codiaceae, 154, 168
Codium taylorii, 154, 168
Combretaceae, 75, 78–79
common cattail, 12, 55
Conocarpus erectus, 75, 78
Convolvulaceae, 3, 12, 13, 33
Corallinaceae, 91, 97
Corallinales, 91, 97
Cyanobacteria, 10, 88
Cymodocea filiformis, 59–63
Cymodoceaceae, 60
Cyperaceae, 12, 13, 44
Cystocloniaceae, 92, 123

INDEX

Dasycladaceae, 155, 173
Dasycladales, 155, 173
Dictyota menstrualis, 138, 140–142
Dictyotaceae, 138, 140
Dictyotales, 138, 140
Digenea simplex, 92, 112–113
Distichlis spicata, 9, 12, 47–48

Ectocarpaceae, 138–139, 149
Ectocarpales, 138, 146
Ectocarpus rallsiae, 139, 149
Ectocarpus siliculosus, 139, 149
Eustoma exaltatum, 12, 35–36,
Evening Primrose Family, 12, 36
evening primrose, 12, 36

Feldmannia indica, 138, 146
fiddleleaf morning-glory, 12, 33
Fimbristylis castanea, 12, 44–45
Fucales, 139, 151

Gelidaceae, 92, 121
Gelidiales, 92, 121
Gelidium crinale, 92, 121–122
Gelidium pusillum, 92, 121
Gentian Family, 12, 35
Gentianaceae, 12, 13, 35
Goatfoot Morning-Glory, 12, 34,
Gracilaria tikvahiae, 92, 128–129
Gracilariaceae, 92, 128
Gracilariales, 92, 128
Grass Family, 46
Grateloupia filicina, 92, 131–132
Grateloupia pterocladina, 92, 131
green seaweeds (algae), 88, 154
gulf cordgrass, 12, 52

Halimeda incrassata, 155, 171
Halodule beaudettei, 59–60, 63–64, 70, 113, 116
Halodule wrightii, 59–60, 63
Halophila engelmannii, 60, 65–66
Halymenia floridana, 92, 133
Halymeniaceae, 92, 131
Halymeniales, 92, 131
Heliotrope Family, 11, 23
Heliotropium angiospermum, 11, 23
Heliotropium curassavicum, 9, 11, 23–24
Heliotropium racemosum, 11, 23–24
Hincksia mitchelliae, 138, 147
Hydrocharitaceae, 60, 65
Hydrocotyle bonariensis, 11, 16, 17
Hydrocotyle umbellata, 11, 16, 17
Hydropuntia caudata, 92, 128, 130
Hydropuntia cornea, 92, 128, 130
Hypnea cornuta, 92, 123, 125
Hypnea musciformis, 92, 123–125
Hypnea spinella, 92, 123–124

Hypnea valentiae, 92, 123–125
Hypneaceae, 92, 123

Ipomoea imperati, 12, 33
Ipomoea pes-caprae, 12, 33–34
Ipomoea sagittata, 12, 33–34

Jania adhaerens, 91, 97
Jania capillacea, 91, 97–100
Jania cubensis, 91, 97–99
Jania subulata, 91, 97, 100

Laguncularia racemosa, 75, 79
Laurencia poiteaui, 91, 115
Leadwort Family, 12, 38
Limonium carolinianum, 9, 12, 38–39, 40
Lomentaria baileyana, 93, 136–137
Lomentariaceae, 93, 136
Lycium carolinianum, 2, 9, 12, 41

Machaeranthera phyllocephala, 9, 11, 18, 19
manatee grass, 60–63
marsh pennywort, 11, 16
mermaid's wine cup, 175
Monanthochloe littoralis, 9, 12, 46, 49–50
Morning-Glory Family, 33
Mustard Family, 11, 25

narrow-leaved cattail, 12, 55
Neosiphonia gorgoniae, 92, 113–114
Neosiphonia tepida, 92, 113
Neptune's shaving brush, 172
Nightshade Family, 12, 41

Ochrophyta, 88, 138, 140
Oenothera drummondii, 12, 37
Onagraceae, 12, 13, 36
Opuntia engelmannii, 11, 27
Opuntia macrorhiza, 11, 27
Opuntia stricta, 11, 27

Padina gymnospora, 86, 138, 143–144
Palisada poiteaui, 92, 115–116
Penicillus capitatus, 86, 155, 172
pennywort, 11
perennial glasswort, 9, 30
Petalonia fascia, 139, 150
Phaeophyceae, 88, 138, 140
Pigweed Family, 11, 28
plains prickly pear, 11
Plumbaginaceae, 12, 13, 38
Poaceae, 12, 13, 46
Polyphysaceae, 155, 175
Polysiphonia atlantica, 92, 117–118
Polysiphonia boldii, 92, 117–120
Polysiphonia denudata, 92, 117–118
Polysiphonia echinata, 92, 117–119
Polysiphonia hapalacantha, 92, 117–119

Polysiphonia havanensis, 92, 117–119
Polysiphonia ramentacea, 92, 117–119
Polysiphonia subtilissima, 92, 117–119
Porphyra rosengurtii, 91, 96–97
Primrose Family, 12, 40
Primulaceae, 12, 13, 40

red mangrove, 75, 80–81
red seaweeds (algae), 88
Rhizophora mangle, 75, 80–81
Rhizophoraceae, 75, 80–81
Rhodomelaceae, 91, 106
Rhodophyta, 88, 91
Rhodymenia pseudopalmata, 93, 135
Rhodymeniaceae, 93, 134
Rhodymeniales, 93, 134
Rhynchospora colorata, 12, 44, 46
Ruppia maritima, 60, 69–71
Ruppiaceae, 60, 69

Salicornia bigelovii, 9, 11, 27–30
Salicornia virginica, 9, 11, 27–30, 49
Salt Cedar Family, 42
salt cedar, 42–43
salt meadow cordgrass, 12, 52
salt-flat grass, 9, 12, 49–50
saltgrass, 9, 12, 47–48
saltmarsh bulrush, 44
saltmarsh fimbristylis, 12, 45
saltmarsh morning-glory, 12, 34
Saltwort Family, 11, 20
saltwort, 9, 11, 20–22, 49
Samolus ebracteatus, 12, 40
Sargassaceae, 139, 151
Sargassum filipendula, 139, 151, 153
Sargassum fluitans, 139, 151–152
Sargassum natans, 139, 151–153
Scirpus maritimus, 12
Scytosiphonaceae, 139, 150
sea blite, 9, 11, 31–32
sea lavender, 9, 12, 38–39
sea lettuce, 156
sea oats, 12, 54
sea ox-eye daisy, 9, 11, 18
sea purslane, 9, 11, 14
sea rocket, 11, 25–26
seabeach pimpernel, 12, 40
seaside heliotrope, 9, 23–24
Sedge Family, 44
Sesuvium maritimum, 11, 14
Sesuvium portulacastrum, 9, 11, 14

Sesuvium trianthemoides, 11, 14, 15
Sesuvium verrucosum, 11, 14, 15
shoal grass, 60, 63–64
smooth cordgrass, 9, 12, 51
Solanaceae, 3, 12, 13, 41
Solieria filiformis, 126–128
Solieriaceae, 92, 126
southern spineless cactus, 11
Spartina alterniflora, 9, 12, 47, 51
Spartina patens, 12, 47, 51–52
Spartina spartinae, 12, 47, 51–52
Spatoglossum schroederi, 138, 145
Sporobolus virginicus, 12, 46, 48, 53
Spyridia filamentosa, 91, 105–106
Spyridia hypnoides, 91, 105–106
Spyridiaceae, 91, 105
Suaeda conferta, 11, 31–32
Suaeda linearis, 9, 12, 31–32
Sunflower Family, 11
Syringodium filiforme, 59

Tamaricaceae, 3, 12, 13, 42
Tamarix aphylla, 12, 42
Tamarix canariensis, 12, 42
Tamarix chinensis, 12, 42–43
Tamarix gallica, 12, 42–43
Tamarix ramosissima, 12, 42–43
Texas prickly pear, 11, 27
Thalassia testudinum, 60, 63, 67–69, 148, 174
turtle grass, 60, 67–69
Typha domingensis, 12, 55
Typha latifolia, 10, 12, 55
Typhaceae, 12, 13, 55

Udoteaceae, 155, 171
Ulva clathrata, 154, 156–157
Ulva fasciata, 154, 156–159
Ulva flexuosa, 154, 156–157
Ulva lactuca, 154, 156–159
Ulva prolifera, 154, 156, 158, 160
Ulvaceae, 154, 156
Ulvales, 154, 156
Ulvophyceae, 154, 156
umbrella pennywort, 11, 16
Uniola paniculata, 12, 46, 54

Virginia dropseed, 12, 53

white mangrove, 75, 79
white-topped sedge, 12, 46
widgeon grass, 60, 69–71

Harte Research Institute for
Gulf of Mexico Studies Series

Coral Reefs of the Southern Gulf of Mexico
edited by John W. Tunnell Jr., Ernesto A. Chávez, and Kim Withers

Gulf of Mexico Origin, Waters, and Biota: Volume 1, Biodiversity
edited by Darryl L. Felder and David K. Camp

Gulf of Mexico Origin, Waters, and Biota: Volume 2, Ocean and Coastal Economy
edited by James C. Cato

Encyclopedia of Texas Seashells
by John W. Tunnell Jr., Jean Andrews, Noe C. Barrera, and Fabio Moretzsohn

Gulf of Mexico Origin, Waters, and Biota: Volume 3, Geology
edited by Noreen A. buster and Charles W. Holmes

Gulf of Mexico Origin, Waters, and Biota: Volume: 4, Ecosystem-Based Management
edited by John W. Day and Alejandro Yáñez-Arancibia